D1544207

DATE DUE

BRODART, CO.   Cat. No. 23-221

ANCIENT BODIES, MODERN LIVES

ANCIENT BODIES, MODERN LIVES

How Evolution Has Shaped
Women's Health

Wenda Trevathan

OXFORD
UNIVERSITY PRESS

2010

OXFORD
UNIVERSITY PRESS

Oxford University Press, Inc., publishes works that further
Oxford University's objective of excellence
in research, scholarship, and education.

Oxford   New York
Auckland   Cape Town   Dar es Salaam   Hong Kong   Karachi
Kuala Lumpur   Madrid   Melbourne   Mexico City   Nairobi
New Delhi   Shanghai   Taipei   Toronto

With offices in
Argentina   Austria   Brazil   Chile   Czech   Republic   France   Greece
Guatemala   Hungary   Italy   Japan   Poland   Portugal   Singapore
South Korea   Switzerland   Thailand   Turkey   Ukraine   Vietnam

Published by Oxford University Press, Inc.
198 Madison Avenue, New York, New York 10016

www.oup.com

Oxford is a registered trademark of Oxford University Press

Library of Congress Cataloging-in-Publication Data
Trevathan, Wenda.
Ancient bodies, modern lives : how evolution has shaped women's
health / Wenda Trevathan.
p.   cm.
Includes bibliographical references and index.
ISBN 978-0-19-538888-6 (hardcover)
1. Women—Health and hygiene.   2. Evolution (Biology)   I. Title.
RA778.T673 2010
613'.04244—dc22
                                                2009039430

1   3   5   7   9   8   6   4   2

Printed in the United States of America
on acid-free paper

Contents

Acknowledgments

Writing a book like this requires a number of key ingredients. One is the body of sophisticated and exciting research on reproductive biology and health from which I have drawn extensively. A quick glance at the list of references cited provides a good compilation of the work that I believe has the most to offer as we try to understand challenges to women's health that we are facing and will continue to face as global resources constrict, population expands, and more and more people strive for the lifestyles of the "health-rich" nations. I want to particularly call attention to some of the scholars whose work I cite frequently and in almost every chapter. They include Virginia Vitzthum, Carol Worthman, Jim McKenna, Gillian Bentley, Barry Bogin, Jim Chisholm, Kathryn Dettwyler, Boyd Eaton, Peter Ellison, Paul Ewald, Dean Falk, Tuck Finch, Helen Fisher, Helen Ball, Kristen Hawkes, Sarah Hrdy, Chris Kuzawa, Jane Lancaster, Lynette Leidy Sievert, Thom McDade, Randy Nesse, Catherine Panter-Brick, Tessa Pollard, Karen Rosenberg, Dan Sellen, Meredith Small, Stephen Stearns, Beverly Strassman, David Tracer, Pat Whitten, and Andrea Wiley. In addition to this being a group of exceptional scholars, I am immensely pleased to say that they are my friends. And as most authors have said, the good stuff in this book is theirs; the mistakes in interpreting their work are mine.

Another important requirement for writing a book is time. Ask any academic (perhaps any professional from any field) what commodity they desire the most and virtually all will answer "time." The information was there in the publications of my colleagues, the ideas were in my head, the benefits of what I wanted to say seemed clear, and the potential audience appeared receptive. What I didn't have was the time to write it all down. For that, I thank the School for Advanced Research (SAR) in Santa Fe, New Mexico. When I received their invitation to spend nine months as a resident scholar on their beautiful campus, I knew that I finally had the opportunity to complete a book that had been spinning around in my mind and in my classrooms for a decade or more. The peace and quiet, the intellectual stimulation from fellow resident scholars and staff, and the beauty of northern New Mexico came together to create the perfect opportunity for reading, writing, and reflection. It is difficult to find the words to express my appreciation to SAR and all who support it for my time there.

I am appreciative of helpful feedback and responses to inquiries from Virginia Vitzthum, Caleb "Tuck" Finch, Kathryn Dettwyler, Chris Kuzawa, Sara Stinson, Mel Konner, and my "consultants" Jennie Lentz, Marcia Trevathan, and the students in my Anthropology of Reproduction class on the Semester at Sea/Fall 2009. I am particularly grateful to two anonymous reviewers who read the manuscript carefully and provided useful and constructive suggestions for improving it. Especially helpful to me in Santa Fe were conversations with Dean Falk, Rebecca Allahyari, Nancy Owen Lewis, Linda Cordell, and Nick Thompson. I thank the editors and staff at Oxford University Press, most especially Sarah Harrington, for their patience as I corresponded with them from a peripatetic itinerary that literally spanned the globe. Finally, for years of support, friendship, advice, and encouragement, I thank Jim McKenna, Carol Worthman, Karen Rosenberg, Mary Burleson, Dean Falk, Neal Smith, Jack Kelso, Earl Trevathan, Sue Trevathan, and last, but far from least, Gregg Henry.

ANCIENT BODIES, MODERN LIVES

Introduction: What Does Evolution Have to Do with Women's Health?

We may live in both the best of times and the worst of times, although I'm sure that Charles Dickens would beg to differ. For those of us who have the fortune to live in health-rich nations like the United States, Canada, Japan, and those of Western Europe, we enjoy a level of health that was unimaginable when Dickens was writing. "Seventy is the new fifty" and not only are we living longer, we are doing so in a state of fairly good health. But for people in health-poor nations, the scenes that Dickens described of London and Paris in the 18th century would be all too familiar. Globalization, which seems to improve standards of living for some people, has had profoundly negative effects in many parts of the world. Political and social scientists tackle some of the big issues that arise from increasing health and economic inequality, and even in the health-rich nations we know we can do better to improve overall health and quality of life.

Scientists who study *evolutionary medicine* attempt to determine, by better understanding evolutionary processes, how some of the health problems we face today arose. This approach will not solve all of our entrenched and deep-rooted health challenges, but it does offer a new, more holistic way of viewing individual and population health from both research and practice standpoints. In this book I hope to describe this approach with regard to the health of women.

One common refrain of evolutionary medicine is that our biological selves are not well matched with our contemporary lives, and the result of this "discordance" is poor health, especially with regard to chronic and degenerative diseases and disorders such as type 2 diabetes, cardiovascular disease, cancers, and hypertension.[1] If we could return to the lifestyles of the Stone Age, the argument runs, we would quickly become healthier. Certainly there are practices from earlier times that would probably improve our health, most notably dietary and exercise modifications, but with more than 6 billion people on the planet, the reality of "returning" to prehistoric ways of living is out of reach for the vast majority of us. Furthermore, one of the most obvious mismatches between our evolved bodies and today's environments results from newly created biochemicals that are in our foods,

3

water, clothing, furniture, and the air we breathe and for which our bodies have not evolved adaptations.[2] There is not much of an evolutionary take on this except to state the obvious (that is, all of these chemicals are products of modern technology), and there is little chance that we will be able to return to the Stone Age as regards several decades of "better living through chemistry."

In my mind, the best that evolutionary medicine can offer is new ways of thinking about old diseases and disorders, particularly those for which medical research has been unable to find a cure or other highly successful treatments. Evolutionary medicine offers a broader and more inclusive approach to medicine and health and urges us to ask new questions about both the immediate and developmental causes and explanations of poor health. Most medical research is highly specialized and focuses on fairly narrow windows into diseases and disorders. This is understandable and necessary because of the continuing expansion of knowledge about disease processes at the molecular, cellular, and organ system levels. Social and behavioral scientists urge concern for behavior, psychology, and the sociopolitical environments when considering not only causes of diseases but treatments as well. Evolutionary medicine steps back further and takes into consideration the entire species and its evolutionary history.

So what does evolution have to do with women's health?

Plain and simple, evolution is about reproduction. And since so much of what women are biologically is about reproduction, almost every book dealing with biology and evolution and women will focus on reproduction—this one is no exception. Whether we reproduce or not, our evolutionary history and our current biology were shaped by natural selection operating to increase reproductive success. This means that almost anything we have to say about women's health from an evolutionary perspective will involve reproduction in one way or another. On the other hand, reproduction and virtually every other aspect of human life take place in dense social and cultural contexts that also shape and define them. For example, *menarche* is not just the first instance of menstruation; in most cultures of the world, it comes with new social roles and status for girls who from that point on may be eligible for marriage, be required to adopt new clothing or hair styles, or be seen as women rather than girls. If we talk only about the biological aspects of menarche, we will lose track of what it means to be human. This is no less true for other aspects of reproduction and women's biology.

As noted, one of the common approaches of evolutionary medicine argues that contemporary health problems to some extent result from a mismatch between our evolved bodies and our present lifestyles (culture).[3] What is usually meant by "evolved bodies" is the physical human form that resulted from millions of years of evolutionary processes at work on our ancestors from the origin of the primate order (approximately 60 million years ago) to the origin of food production (agriculture) approximately 10,000 years ago. This is not to imply that evolutionary processes stopped 10,000 years ago, but it does highlight the extraordinarily short time during which the pace of cultural evolution has far outstripped the pace of biological evolution. In this view, 10,000 years ago, a mere blip in the time span of evolution, marked the beginning of a veritable explosion of culture and technology.

For most of the span of human existence, individuals were born into and lived their lives in pretty much the same environment. Furthermore, they experienced life in environments not very different from those of their recent ancestors. Thus, it was not necessary for them to respond to significant environmental changes within one or several lifetimes. Today, however, with mass migrations, major environmental challenges, and unavoidable sociopolitical changes occurring within single lifetimes, many humans are stressed to their limits (and often beyond) to adapt physically, emotionally, and materially, and to do so in ways that often result in considerable challenges to health, especially reproductive health. Not only are we facing a "third epidemiological transition" with the reemergence of infectious disease[4] but we are also facing an epidemiological *collision* as the health challenges of the past (high infant mortality, malnutrition, infectious disease) meet those of the present (obesity, cardiovascular diseases, cancer) in ways that will severely test our abilities to respond.

In fact, even though our ancestors may not have had to face major environmental changes in their lifetimes, natural selection has favored in the human species a level of adaptability (also known as *plasticity*) that is unusual in the mammalian world. For example, although we are *omnivores* (that is, we can eat all kinds of things and lack a specialized digestive system), we can also adapt to diets very high in animal foods (such as a traditional Arctic Inuit diet) or become exclusively vegetarian, depending on habitat and habits. Of course, there are limits to our adaptability as would be quickly seen for someone whose chosen or imposed diet excluded essential nutrients like ascorbic acid (vitamin C) or vitamin B-12. Perhaps few aspects of human life have changed as much as diet, and few aspects have as great an impact on health as diet. And as we will see, reproduction has a lot to do with food.

Throughout this book I will talk about "our ancestors" or "ancestral humans." As noted earlier, the context of human lives changed considerably with the beginnings of food production (agriculture and animal domestication) less than 10,000 years ago. Human reproductive biology was shaped during the long period from the origin of the human lineage about 5 to 7 million years ago until the present, but almost all the time during which natural selection operated on reproduction occurred before the advent of agriculture. This period is often referred to as the environment of evolutionary adaptation or EEA, and when we discuss the concept of mismatched bodies and lives, we usually mean the body that was shaped during that period. When I discuss ancestral humans in this book I usually refer to those who lived in that bulk of evolutionary history. Unfortunately, most reproductive variables and behaviors do not leave their marks in the fossil record, so I and others working in the area of evolutionary medicine must rely on studies of living non-human primates and human populations who lead existences that we think resemble somewhat those of our ancestors.

For example, studies of our closest living relatives—chimpanzees, gorillas, and orangutans—reveal that females in captivity reach puberty earlier, give birth first at younger ages, and have lower birth intervals than those in the wild.[5] In most cases, captive great apes have more reliable and more abundant food and they are less active, mirroring in some ways the changes that occurred when humans began producing their own food and living in settled communities. (Perhaps it can be said that captive apes and food-producing humans have become "domesticated," and that they experienced changes in reproductive biology similar to those experienced by other domesticated animals.) Thus, to the extent that there are similarities in the reproductive physiology of apes and humans, comparative research on wild and captive populations can provide insights into ancestral human biology and behavior and the changes that occurred with domestication.

Of course, chimpanzees and gorillas are not ancestral humans. Neither are the foraging populations of Africa and South America that have often been used as "proxies" for our ancestors because their ways of living are presumably somewhat like those of our ancestors, at least in comparison to the lives of 21st-century men and women in industrialized nations. But these are only rough proxies and much of the discussion of "ancestral bodies" must remain at the level of conjecture, albeit conjecture based on pretty good reasoning.

Often when one uses the term *conjecture* to refer to an idea, it has a derogatory connotation that suggests that the idea is just somebody's wild, unfounded notion. For those of us who provide evolutionary explanations, however, it is an admission that we can never know for certain that we are right, we can never "prove" anything about human thought and behavior that occurred far in the past. (Actually, proof is never a goal of science; support or refutation of a hypothesis is the most one can strive for because the scientific method requires that every explanation be open to rejection with new data or new interpretations of existing data.) But that does not mean that we are not good scientists. As with any scientific explanation, we put forth a possible scenario and it stands or falls based on how well it explains what it proposes to explain. New fossil evidence, new observations about nonhuman primates, new data based on comparative DNA analysis, even new "wild ideas" can alter or outright reject any scenario. This is how science works and how knowledge proceeds, including aspects of science that explore ancestral biology and behaviors.

As I describe some of the proposals about ancestral humans throughout the book, I often present the ones that make the most sense to me or seem to have more "explanatory power." In some cases I provide several alternative views, but in others, one explanation may seem so strong to me that I focus almost exclusively on that one. Readers should keep in mind, however, that there are almost always several ways of viewing a given phenomenon and if I were to write a second edition of this book in a few years, chances are good that some of the explanations would need to be altered significantly because of new information.

Evolutionary Medicine

Defined simply, evolutionary or Darwinian medicine is the application of principles of evolutionary theory to human health and disease (and, we hope, to medical practice and research). Although most recent discussions of the field trace its origin to works in the 1990s by evolutionary biologists Randy Nesse and George Williams,[6] the usefulness of an evolutionary lens for viewing human health has been recognized for at least a hundred years.[7] Unfortunately, that does not mean that an evolutionary perspective is currently being embraced by medical practice, which would be a key measure of its usefulness.[8] Nesse and Stephen Stearns, another evolutionary biologist,[9] cite a number of reasons that evolutionary theory is not incorporated into medical research and practice, including the scarcity of evolutionary biologists on medical school faculties, the very little exposure that medical students get to evolutionary thinking, and the fact that many medical students (in the United States, at least) do not accept evolution as the basis of biological sciences. Given the complexity and burden of the current medical curriculum, it is unlikely that courses in evolutionary medicine will be added any time soon. This is unfortunate, because it actually might lighten their burden if students were armed with understanding of theories rather than the endless details that are deemed necessary to pass medical boards. As Nesse and Stearns note, evolutionary biology "can help make medical education more coherent by giving students a framework for organizing the required 10,000 facts."[10]

Another point at which medicine and evolutionary biology are sometimes at odds is in their views of the body. A common metaphor found in medicine is that the body is a machine designed for certain functions and when things go wrong, it can be treated in much the same way that an auto mechanic treats a poorly functioning car. Add a little oil here, tighten the belts, clean the carburetor, replace the spark plugs, adjust the tire pressure, or remove the possum from the engine compartment. The machine was designed by an engineer using blueprints that can be consulted when making the repairs. The human body, however, is a bundle of "compromises shaped by natural selection in small increments to maximize reproduction, not health."[11] This leaves us vulnerable to lots of diseases and disorders, but it also makes us amazingly resilient. Nesse and Stearns suggest that if medicine would give up the idea that the body is a machine, physicians would find it easier to place the discipline on firmer biological foundations. Evolutionary medicine attempts to move medical research and practice in that direction.[12]

For example, an evolutionary view of infectious disease can add a great deal to our understanding of newly emerging diseases and may enable epidemiologists to predict where an outbreak will occur and how it will progress.[13] The evolutionary view warns us that trying to treat bacterial infections with ever more powerful antibiotics will ultimately backfire as the bacteria evolve resistance to our medical interventions. It may also help us deal with major health crises of our time such as the human immunodeficiency virus (HIV) by figuring out a way for evolution to work *for* us in directing the course of microbial evolution from greater to less virulence, rather than the usual arms race we employ that results in greater and greater virulence.[14] We may be able to develop successful vaccines against a number of diseases if we have a clear picture of how they have evolved. Because both clinical medicine and evolutionary medicine have as their goals the control of diseases and the minimization of misery, the evolutionary perspective has hope of slowly becoming part of research and practice. The money, time, and cost of mistakes makes attempting to completely eradicate a disease a poor use of resources and one that will likely fail. Leading pathogens to a state of benign coexistence through evolutionary mechanisms would be a better use of resources and intellectual talents and would promise much greater success. This means we would have a nice barnyard of domesticated critters rather than a forest of dangerous ones with the potential to kill.

As in many other modern health movements, consumers are often the engine that drives change by requesting that their physicians consider, for example, gender, inequality, or lifestyles in their treatments and health recommendations. Feminist writing and persuasion introduced into medicine new ways of viewing childbirth, reproductive technologies, human subjects in medical research, and reproductive cancers.[15] The language used in medical discourse has undergone change (perhaps not enough) in the past few decades because of feminist critiques.[16] Anthropology has attempted to make medicine aware of the great amount of variability that exists in the human species, including variability in "normal health."[17] Perhaps among the readers of this book will be health consumers who make possible an expansion of the medical view to include evolutionary as well as sociocultural factors in treatments and recommendations. As noted, scholars writing in the field have so far had limited success with facilitating acceptance of evolutionary principles in medicine.

Evolutionary Theory

A number of fundamental concepts from evolutionary theory underlie much of the work in the field of evolutionary medicine and are discussed throughout this book. I assume that most readers have a general understanding of how the evolutionary process works: natural selection operates on traits, behaviors, and characteristics that promote health, survival, and, most important, reproductive success, in a given environmental context—and against those that compromise health and reproductive success. In order for any characteristic to have an evolutionary basis, it must have some underlying genetic basis, it must vary, and it must be heritable. This last requirement is where reproduction comes in. The only thing that "counts" as far as the process of evolution is concerned is whether or not a trait is passed on to succeeding generations, and the only measure of evolutionary success is reproductive success (also known as *fitness*). For instance, if a woman has a genotype that results in sterility, she will not pass that trait along to her direct descendents (because she will not have any), so the trait will not evolve. In the language of evolution, the trait has a negative effect on fitness and will not persist. On the other hand, a trait that enhances reproductive success will have a positive effect on fitness and will be passed down through generations. But, as we will see, whether a trait is positive or negative for reproductive success depends almost entirely on context.

Evolution has often been described as a "selfish" process. According to this view, individuals (actually, individual genes) will do almost anything to get their genes into the next generation. The result is intense competition between individuals to get ahead in the evolutionary race. The individual whose genes are represented in the greatest numbers several generations down the line is, at least for a time, the "winner." To get those genes into succeeding generations, individuals have *reproductive strategies* to find the best mate (known as *mating effort*), provide the best care of offspring (*parental effort*), and help others with shared genes do the same (*kin selection* or *inclusive fitness*). The strategies lead to allocation of time, energy, and other resources toward the goal of increased reproductive success. Many of the terms used to describe behaviors related to reproduction are not ideal (like *selfish, competition, strategy, goal, allocation*) because they imply conscious intent, but the behaviors and traits are simply the results of whatever characteristics lead to greater numbers of surviving genes in subsequent generations.

Another somewhat misunderstood concept that plays an important role in evolution and in the topics covered in this book is that of *parent-infant conflict*. Most of us have intimate experience of this conflict in the colloquial sense if we were ever teenagers or had teenage children. But when it is used in an evolutionary sense this concept refers to the fact that the interests of two individuals are never the same, and when the needs of one compromise the health or reproductive success of the other, conflict in one form or another often results. Consider, for example, that a mother (who is related to her offspring by about 50 percent of her genes) will do everything she can (theoretically) to enhance the survival of a given child as long as it does not interfere with the survival of other young she has or can potentially have. The child is related to himself by 100 percent of his genes, so he, in turn, will do everything he can to survive, even if it compromises the health of his mother or current and future siblings. As we will see in later chapters, this comes into play in pregnancy and during breastfeeding when the needs of the developing fetus or nursing infant and of the mother are not always the same. The infant may try to prolong the free ride and nurse as long as he can, but the mother needs to terminate breastfeeding in order to reproduce again and to preserve her own energy reserves.

Proximate and Ultimate Causation

When we examine a characteristic that influences health, survival, and reproduction, it is important to consider that "causes" can be both *proximate* and *ultimate*. The proximate cause is usually the immediate one and the one that a clinician might write on the medical chart. The proximate cause is what is most commonly treated by drugs, surgery, or other medical procedures. A frequently cited example is that of a fever in response to an infectious agent. The cause of the fever is a virus or bacterium and a treatment would involve seeking something to wipe out the cause or reduce the symptoms. But an evolutionary, or ultimate, perspective argues that a fever is one of the best tools the body has to fight off infection, a tool that has evolved as part of a complex immune system that enhanced survival and reproductive success in the past.[18] In fact, it may be one of the few tools at our disposal for dealing with viruses, which do not respond to antibiotics. In the language of evolutionary medicine, a fever is a defense rather than a defect.[19] Of course, like any defense mechanism, it can get out of control and become a problem that needs medical intervention. Certainly a fever of 105°F in a 3-year-old child would not be considered a healthy response to an infection and would require treatment to bring it down.

Another example comes from considering why a person develops sickle cell anemia. The proximate focus may be on the genetic anomaly that caused the anemia: people with two alleles[20] for sickle cell (hemoglobin S rather than the "normal" hemoglobin A that most people have) usually develop a severe form of anemia that most commonly resulted in death or severely compromised health before age 20 in the past, as it often does today in the absence of medical intervention. When a person seeks medical help for anemia, most clinicians seek the proximate cause or the specific mechanism that caused the disorder in the individual and this is the focus of treatment.

On the other hand, concern for the ultimate explanation for the high prevalence of the sickle cell allele would lead to assessing why the alleles are distributed in such a way that they reach high frequencies among those living in or descended from people in Africa and the Mediterranean. What selective value does the sickling allele have in these regions that would explain the high frequencies? Anthropologist Frank Livingstone,[21] who can be considered one of the first scholars of evolutionary medicine (although it was not called that at the time he was writing), noted that the geographic distribution of the sickle cell allele coincides with the distribution of malaria. His conclusion was that those who had a single allele for sickle cell (referred to as "sickle cell trait") were less likely to develop or die from malaria than those who had two copies of the allele for "normal" hemoglobin A. Thus, those who were heterozygous for hemoglobin S had greater reproductive success in malarial areas, and the allele is maintained in high frequencies even today. This hypothesis, that sickle cell trait is protective against malaria, has been tested and confirmed, and although knowing this has not necessarily been useful in treating sickle cell anemia, it has led to greater understanding of the "adaptive significance" of the allele and its pattern of distribution.

Another way of thinking about proximate and ultimate causation is to consider that proximate causes are things that happen to an individual in a lifetime, whereas ultimate causes affect populations and species over much longer spans of time, thousands rather than dozens of years. This is a key to knowing why clinical medicine has been slow to embrace evolutionary medicine—clinicians who focus on the ultimate or evolutionary explanation (who "practice" evolutionary medicine) probably will not have as much impact on their clients' health and well-being as those who focus on and treat the proximate causes.

Evolutionary biologist Paul Ewald has expressed concern that when evolutionary medicine tries to develop ultimate or evolutionary explanations for diseases and disorders based on well-established proximate medical explanations, the wrong explanation may result if the proximate medical causes are themselves wrong.[22] The risk of not being right about evolutionary explanations increases when infectious causes are not considered, and Ewald has been critical of much of evolutionary medicine for ignoring them. For example, when the assumed cause of peptic ulcers was stress interacting with a particular genotype, an evolutionary explanation based on that assumption would have been wrong once it was determined that ulcers had infectious origins.

If our focus on "causes" of cancers is restricted to mismatches between evolved biology and 21st-century lives, we may ignore evidence that infectious agents play roles. Ewald argues that evolutionary medicine approaches to every disease and disorder should consider all three categories of explanation: genetic, infectious, and environmental/lifestyle.[23] This is particularly important in investigations of causes when treatment and prevention are the goals of research. Clearly, treating a disease that is presumed to be caused solely by genes but is, in fact, infectious in origin, will result in failure or incomplete healing if the infectious agent is not considered. Telling people to relax and remove stress from their lives did not really cure them of ulcers caused by a bacterium (although it was probably helpful for their overall health). On the other hand, we know that tuberculosis is cause by a bacterium, but how it is manifested and how it is treated depend a great deal on both genes and environment (such as diet and hygiene). If it turns out that some severe cases of morning sickness (see Chapter 4) are caused by pathogens, making a case that the nausea is adaptive because it "protects the embryo" may not only backfire but could also result in the death of the mother and her fetus from severe dehydration. In this case, the watchful waiting recommended by evolutionary medicine would be the wrong prescription. If evolutionary medicine is to have any impact at all, it must live up to its claim of taking a holistic and deep-time view of human health.

The Concept of "Normal"

Those who view human health and development through the lens of evolutionary medicine take as their beginning assumption the idea that humans are highly diverse with a biology that unfolds in a context rather than in predictable, unvarying ways. Modern medicine tends to think in terms of "normal" and averages, with the model for normal usually being well-nourished people in health-rich populations. Clinicians tend to treat the deviations from normal, but what is normal for women growing up in relative affluence may be very different for women in many populations today and in the ancestral past. Some of the perceived deviations from normal may actually be healthful responses to local circumstances. Evolutionary medicine recognizes that for almost any human trait, there is a broad range within which growth and development can occur and living and reproducing can be successfully maintained. We will see throughout this book, however, that deviations when circumstances require it are often examples of the body behaving in a "good enough" way to survive and reproduce.

Human Evolution and Life History

Living requires energy. Growing requires energy. Reproducing requires energy. Healing requires energy. Energy, in the form of calories from the food we eat, does not seem to be in short supply to most of us who are reading this book, but for animals living in the wild and for humans throughout evolutionary history and in some regions today, energy is not only in limited supply but availability also varies greatly throughout a year or a lifetime. Thus, natural selection has favored mechanisms that work to allocate energy among bodily functions so that the end result for an individual (or kin group) is increased reproductive success. Because energy is limited, this allocation process requires *trade-offs*, an important concept in life history theory.[24] In short, unless resources are abundant and the work required to obtain them is low, women cannot have well-functioning immune systems, fat babies, and healthy toddlers all at the same time. Something has to give; one has to be "traded" for another. Life history theory helps predict what will be traded for what, under certain environmental conditions, including social and cultural circumstances.[25]

The concept of trade-offs also informs us that reproductive success is not just about quantity but, especially for humans, it is equally or more importantly about quality. It is not just a matter of producing lots of offspring, but producing offspring that survive and reproduce. Avian ecologists long ago demonstrated that reproductive success (via survival and next-generation reproduction) was often greater when there were fewer nestlings and that very few birds reproduce at the maximum.[26] Imagine a woman who has 10 children but none of them survives and one who has only a single child who lives to have several grandchildren. Which has the greatest reproductive success? For humans, the quantity-quality trade-off point is impacted by offspring size, delayed growth, and social sharing of resources and child care. In fact, the practice of sharing may well be the most important factor in life-history strategies for humans. Because of sharing, humans have a higher trade-off point (more quantity) in comparison with other large primates.[27] Here again, culture and the practice of sharing resources makes human life history hard to predict from models created for other species. As anthropologist Robert Walker and his colleagues note, "Humans as large, long-lived, cooperative mammals that reproduce at a fast rate create a formidable life-history combination that probably figured prominently in the successful colonization of hunter-gatherers around the globe."[28]

If an animal is getting sufficient food to maintain life, it is in energy balance: what goes in equals what goes out. An excess of calories that can be used for something else (growth, reproduction, healing) would yield *positive energy balance* (what goes in is greater than what goes out), and a deficit may yield *negative energy balance* (what goes in is less that what goes out). In this last case, the only way that an individual can survive on fewer calories than needed to live is to draw upon energy reserves in fat, muscle, and, ultimately, organs. If negative energy balance is prolonged by ongoing deprivation, death may result. In the short term, however, other energy-hungry functions are put on hold, including growth in children and reproduction in adults. As we will see, women's reproductive functioning seems to be more sensitive to energy balance than men's, but sperm production is also negatively affected by prolonged negative energy balance. As noted earlier, at its most basic level, the study of reproduction is the study of food.

Negative energy balance can interfere with reproduction at several points. Early on in the life course, it can delay puberty and inhibit ovulation following puberty. Lactational amenorrhea (cessation of ovulation because of breastfeeding) is prolonged with negative energy balance and shortened with positive energy balance, a phenomenon that explains why women who are well fed (and thus usually in positive energy balance) cannot rely on breastfeeding to inhibit ovulation. For example, anthropologists Claudia Valeggia and Peter Ellison report that Toba women in Argentina who are well nourished and breastfeed intensively return to positive energy balance soon after giving birth and resume ovulating in less than a year, providing support for the view that energy balance has more of an impact on ovulation than nursing frequency.[29] Energy balance influences hormone levels during menstrual cycling and pregnancy, the timing of births, and infant birth weight.[30] Finally, it appears that prolonged negative energy balance can result in early menopause, although there is still some controversy about this.[31] And, of course, starving people often have decreased sex drives and may not have the energy for courtship and mating activities, further reducing reproductive success.

Many aspects of life history theory can be applied in similar ways to all sexually reproducing animals, but there are a number of characteristics of human life history that make it somewhat unique, at least as we humans see them. In the language of evolution, these unique characteristics are referred to as *derived traits* and serve to distinguish humans from our closest relatives, the great apes. They include longer life spans, later age at first birth, longer growth period, larger maternal body size, larger and more dependent neonates, earlier age at weaning, shorter interbirth intervals, and a prolonged postreproductive life.[32] Also, as anthropologist Gillian Bentley notes, humans today and in the recent past have the ability to "maximize reproduction through cultural interventions,"[33] a characteristic not found in other animals, at least to the degree it exists in humans. Next I will review defining characteristics of humans, including those linked to life history, which may be useful for understanding the arguments presented in the rest of the book.

Bipedalism

In the fossil record, what usually serves as a marker of Homininae, the taxonomic group that includes humans and their immediate ancestors from the genus *Homo* plus australopithecines, is skeletal evidence of bipedalism. This unique character- istic appeared between 5 and 7 million years ago. Although it has been difficult to find an initial "cause" for the evolution of bipedalism,[34] it is clear from a number of experimental studies that it is an energetically conservative way of moving. There must have been many original costs to adopting bipedalism, but the benefit, at least for the last several million years, has been energy efficiency, a critical measure of whether a trait will persist in a lineage or species.

Evolutionary medicine has a lot to say about the medical consequences of bipedalism. Virtually every part of our body had to change for us to become efficient bipeds, and many of these changes have an impact on our health today. Here are some of the things scholars have had to say about bipedalism: "Rube Goldberg" scheme[35]; "bipedalism is a ... bizarre form of locomotion"[36]; "If God is the designer, and we are created in his image, does that mean he has back problems too?"[37] In 1951, Wilton Krogman published a paper in *Scientific American* entitled "The Scars of Human Evolution"[38] that focused primarily on the "scars" resulting from bipedalism. In fact, this paper was one of the first written explicitly about the consequences for health of aspects of human evolution.

Despite these somewhat negative comments, there are a lot of "good things" linked to bipedalism that help define our species today and contribute to our relative success. They include expanded brain size and intelligence; tool using and making; hunting and being able to cover great distances; carrying tools and babies; standing to gather food in trees and bushes; dietary changes toward increased animal consumption; hairlessness and profuse sweat glands for efficient cooling; and, as noted earlier, increased energy efficiency. Now take a look at this list of some of the "bad things" that may be due to bipedalism: pelvic organ prolapse; blood clots and varicose veins; fecal and urinary incontinence; middle ear infections; slipped disks, spondylolysis, and scoliosis; hemorrhoids and hernias; osteoporosis and osteopenia; greater likelihood of choking; plantar fasciitis; and everyday backache, knee pro- blems, and bunions. In fact, several medical specialties owe their existence to bipedalism, including orthopedics, podiatry, physical therapy, chiropractic, osteo- pathy, sports medicine, and obstetrics. This last field, and the impact of bipedalism on pregnancy and childbirth, will be discussed extensively in Chapters 4 and 5.

Huge Brains

All animals have to balance energy allocation among living, growing, staying healthy, and reproducing. Metabolic needs of animals the size of humans are sign- ificant, but body size alone is not the determining factor: humans have metaboli- cally expensive brains. Brains are expensive for all animals, but because humans have such large brains for their body size, the costs are greater than predicted from body size alone. Expanded brain size is another important marker of humans, but it does not appear in the fossil record until about 2–2.5 million years ago with the genus *Homo*. Along with energy efficiency and constraints of the bipedal pelvis on childbirth, brain expansion accounts for an important feature of human life history: an infant born helpless, highly dependent, and requiring prolonged care from mothers and others over many years. In fact, the underdeveloped (at birth) and slow-growing brain is probably responsible for the overall lengthening of the life span and the slowed pace of maturation that characterize our species.[39] The medical consequences of large brains are further discussed in several chapters.

Prolonged Infant and Child Development

A major difference in life history for humans and our close relatives is that all phases are longer for us just as our life span is longer. In fact, when primates are compared (see Figure I-1) for life-cycle phases, they seem to be more similar at the early stages (such as gestation and infancy), but the differences become greater as they move toward total life span. Longer time in a particular stage means more time to grow and learn but also means higher energy demand and delayed reproduction, so there must be some selective advantage to lengthening life-cycle stages. More time to grow also means larger body size, which requires more resources to maintain. Even in the past, humans lived longer than wild or captive chimpanzees and they appear to age more slowly in the adult years.[40]

As noted, a lot of the arguments for why life phases have lengthened in humans revolve around the dependent, slow-maturing human offspring.[41] This helpless little child has been cited as the reason for the human pair bond, why males provision their offspring, our delayed reproduction, low reproductive output, and menopause. At the base of all of this is the fact that raising a child is extremely costly in terms of energy and time and it is difficult, if not impossible, for one parent to do this alone. Successful raising of a child is greatly advantaged if the mother can find help from someone else. Anthropologist Sarah Hrdy has proposed that humans, with their long lives, slowed growth, and highly dependent offspring are cooperative breeders, not unlike those seen in some species of birds and primates.[42] The most successful mothers are those who can turn to *allomothers* (people who can provide care like mothers, no matter their gender) to assist them in resource acquisition and child care. Like most primates, humans are social creatures and we spend our lives with extended networks of kin, so there are usually plenty of others around to share in raising children.

Who is the best candidate for providing this help? If sisters or other reproductive-aged females help, they may jeopardize their own reproductive opportunities. Older children can often provide a great deal of babysitting and other forms of assistance, and doing so gives girls especially valuable experience in taking care of children. Husbands and fathers can be good providers and can make the difference

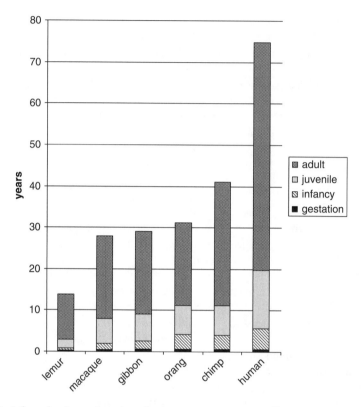

Figure I-1 Life cycle phases for selected primate species. From Jolly, 1985, page 292.

between life and death in many circumstances, but they are not as reliable as Victorian sensibilities and Christian "family values" would have them be.[43] Most anthropological research has found that men have so many other concerns that their children are often of secondary importance. For example, acquiring resources to share with the group at large as a way of enhancing prestige and attracting other mates often takes precedence.[44] Even in the United States, where our ideals of fatherhood are more similar to those of Victorians, as many as 40 percent of children are not living with their genetic fathers.[45] As Hrdy notes, surely we cannot expect more out of ancestral fathers who were at greater risk of early death.

But fortunately, our long lives provide us with a group of people who have everything to gain from providing care for young children and almost nothing to lose: grandmothers. Some scholars have even argued that helpful grandmothers are the key to the success of the human species. For example, Hrdy[46] argues that cooperative breeding, especially involving grandmothers, was the key that enabled children to grow up slowly and thus have time to learn all the things they needed to know to be successful adults and parents. Grandparental provisioning of high-quality foods may have also enabled selection to favor larger and more expensive-to-support brains. The significance of the grandmother, especially if she is beyond the age of producing her own offspring but still healthy enough to provision her grandchildren, will be discussed more fully in Chapters 9 and 10.

Culture

Humans are cultural animals, meaning that we live our lives in the midst of culture and cannot talk about humans without talking about culture any more than we can talk about fish without talking about water. Culture is basically everything about our lives that is not our biology, but, of course, biology influences culture and vice versa. In fact, behavior and culture are often used synonymously because they are so tightly intertwined.[47] When we talk about reproductive strategies in the context of evolution and life history, we often have problems with the mechanistic-sounding terminology because so much of what we do with regard to mating and parenting is under conscious control, shaped by growing up in a specific culture. Even the definition of food is culturally constrained (consider that dog is food in some cultures, emphatically nonfood in others), so it is not surprising that behaviors as basic as mating, birthing, breastfeeding, and parenting are also culturally influenced.

Evolutionary processes have shaped our bodies, but how healthy we are and what diseases and disorders we face in our lives are heavily influenced by the cultural milieu into which we are born and live our lives. For women, whose biology is shaped by natural selection acting on reproduction, our culture influences whether we get married, who we marry, how many sexual partners and mates we have, whether we have children, how many children we have, whether we breastfeed and for how long, and how we perceive and experience menopause. All of these processes related to reproduction can be described biologically, but if our analysis stops there, we will miss entirely what it means to be human. Throughout this book we will discuss how our reproductive bodies have evolved, but we cannot lose track of cultural influences. Even when it is not explicitly stated with regard to a behavior or feature, it is crucial that we be mindful of its all-important role in everything we do.

Hormones

It is impossible to talk about reproduction without having an understanding of the hormones that drive the process in mammals, including humans.[48] In the classic definition, hormones are substances that are produced in one cell, have an influence on other cells, and have wide-reaching, complex effects. Furthermore, their actions can have different effects at different times in the life course, a phenomenon known as *pleiotropy*. When they have positive effects at one point in the life cycle and negative effects at another or under certain conditions, this is referred to as *antagonistic pleiotropy*. For example, estrogens[49] are implicated in both successful reproduction and in reproductive cancers. So it is overly simplistic to say that a specific hormone causes a specific action because of the multiple effects all hormones have.

Just above the roof of your mouth are two of the most important organs related to reproduction: the *hypothalamus* and the *pituitary gland*. They are connected to each other and are in almost constant communication. The hypothalamus has been described as the central command center for all kinds of actions that are going on in the body. The pituitary, on the other hand, is the primary regulator of hormonal interactions related to reproduction. The hypothalamus "tells" the pituitary what to do based on input it receives from throughout the body and brain. There are two parts of the pituitary, the anterior and the posterior. The anterior pituitary secretes the hormones oxytocin and vasopressin and the posterior secretes several hormones that, in turn, act on the gonads (ovaries and testes) and the thyroid, adrenal, and mammary glands.

Hormones can be placed into two major categories: peptide and steroid. Peptide hormones (like prolactin, growth hormone, and oxytocin) are proteins, bind to cell surfaces, and are fast acting because they act on enzymes that are already in the target cells. Steroid hormones (like estrogen, testosterone, and cortisol) are lipids, bind to receptors in the cells, and are slow acting because they direct the synthesis of new enzymes in the target cells. Another difference between the two that has proved important for field research is that peptides are detectible only in blood whereas steroid hormones are detectible in saliva and urine, in addition to blood. This means that hormones can be assessed from saliva, which can be collected under field conditions in a variety of settings where there is no access to refrigeration, clinics, and high-tech labs. Anthropologists have especially embraced the use of saliva as a way of assessing reproductive status under conditions that are arguably more representative of ancestral conditions.[50] Fortunately, because peptide hormones are also important indicators of reproductive and health status, human biologists interested in collecting these under field conditions have developed techniques for assaying dried blood spots, which can be collected from a simple finger prick.[51]

Gonadal hormones are often thought of as male or female hormones. For example, estrogen, produced by the ovaries, is typically referred to as a female hormone and testosterone, produced by the testes, as a male hormone. But ovaries also produce some testosterone and testes produce estrogen, although the quantities are smaller. Furthermore, testosterone can be converted to estrogen. Both have important roles to play in growth, development, and reproduction for both males and females.

The gonadal hormone-producing glands (the ovaries and testes) respond to signals from the pituitary that tell the glands to increase or decrease production of hormones, or cease altogether. Those that target the reproductive organs are called *gonadotropins* and include FSH (follicle stimulating hormone) and LH (luteinizing hormone), which are themselves dependent on signals from gonadotropin-releasing hormone (GnRH) originating in the hypothalamus. FSH and LH influence both ovulation and sperm production; details of these hormones and their actions are discussed in several chapters.

The Anthropological Perspective in Evolutionary Medicine

The field of evolutionary medicine is quite eclectic, comprised of scholars from a number of disciplines, including population and molecular biology, anthropology, psychology, genetics, epidemiology, and clinical medicine. Not surprisingly, the approaches vary according to educational background and theoretical perspectives. Some approaches emphasize clinical applications and others focus on research. Still others try to enhance understanding of contemporary health challenges with concern about how different our lives are from those of our ancestors.

Anthropology depends on cross-species and cross-population research for its conclusions and approaches, most especially on observations and other forms of data based on months and even years of study. Controlled experimentation, the gold standard of medical research, is rarely employed by anthropologists. For this reason, what anthropological writings in evolutionary medicine have to say is often of more interest to clinicians and medical consumers than those conducting basic medical research. As noted earlier, one of the most important contributions to evolutionary medicine (and medicine in general) from anthropology is broadening the definition of "normal" based on observations of the great variation in healthy human biology.

I write as an anthropologist, and most of the research I cite in this book comes from the fields of medical anthropology and human biology. Many of the ideas presented are still at the exploratory or hypothesis-testing stage of knowledge, and the factors affecting reproduction in women are far more numerous and complicated than some of the discussions may indicate. A thorough review of competing hypotheses is beyond the scope of this book, however, and as I noted earlier, I have given more weight to ones that make sense to me than a more scholarly approach would take.

Some readers may be put off by how much I advocate such behaviors as having supportive companionship at birth, allowing mothers and infants to be together for at least the first hour after birth, and breastfeeding for at least a year. My reading of the anthropological and human biology literature and my understanding of human evolution have influenced my thinking on many of the issues that women face today with bodies evolved for reproducing (whether they become mothers or not) and I find that I cannot maintain an unbiased stance on some of them. I also explicitly relate some of the ideas about mismatches to global health and issues of poverty and social justice. This is what I call the "so what?" approach to evolutionary medicine, and it forms the core of the final chapter, where I try to link the research discussed in the previous chapters to health issues that we will face in the future in all parts of the world.

A Note on Terminology

All of us who discuss health in our ancestors and in contemporary populations struggle with what to call people and nations where infant mortality is low, medical resources are widely available, lifestyles are comfortable, food is abundant, physical work is low, birth rates are low, and life expectancy is high. Examples of terms used include *first world* (contra third world), *industrialized* (contra nonindustrialized or traditional), *developed* (contra developing or underdeveloped), *rich* or *affluent* (contra poor), and, most imprecise, *Western* (contra non-Western). Thankfully, terms like *primitive* and *advanced* no longer have currency. For the purposes of this book, however, terminology that emphasizes health seems most useful, so throughout the book I will talk about *health-rich* and *health-poor* populations or nations.[52] This terminology is also useful in reminding us that health may vary extensively within a nation. Thus, when we talk about Western or developed or industrialized nations, that glosses over the populations within a country like the United States for which health statistics look like those in non-Western or nondeveloped or nonindustrialized nations.

There will be times throughout the book when I discuss characteristics of humans, taking a species-wide perspective. It is absolutely crucial, however, to remember that we are a highly diverse, highly adaptable, highly flexible species and that there are very few behaviors or even physiological measures that characterize the entire species, short of bipedalism, a large brain, and a digestive system adapted for a wide variety of foods. At the other extreme, when I refer to an infant or child, I will often use the pronoun "he" to distinguish between mother (always a "she") and child. I fully recognize that there are girl children, but the continuous use of "he or she" can be tiresome.

In popular discourse, we use the term *fertility* to mean anything related to having children or the ability to reproduce. Technically, fertility means the actual production of offspring, and *fecundity* is the more appropriate term for having the potential to bear offspring. A 22-year-old woman with healthy ovarian function but no children is fecund, but not fertile; when she has children, she can be both fecund and fertile as long as she is experiencing ovulatory cycling. If cycles are too short or too long or show no evidence of ovulation, they may be referred to as "subfecund." We usually use the term *infertile* to refer to a woman who is unsuccessful in attempts to conceive, although any woman without children would technically be infertile.

Organization of the Book

This book is organized along the lines of the female reproductive cycle, beginning with the onset of reproductive functioning at menarche and concluding with its termination at menopause. A penultimate chapter addresses our unusual longevity, followed by a chapter that attempts to pull together the material covered in the preceding chapters to provide not only a summary but perhaps a way of thinking about future health, especially with regard to what I call the epidemiological collision.

In the first chapter I focus on the factors that affect the developing reproductive system in girls, noting that the age of onset of ovarian function has decreased in many populations with overall improvements in health over the last 100 years. Early menarche, because it is associated with positive health measures, is often taken to be a good sign, but there is increasing evidence that a price may be paid in later health. One of the consequences of early puberty is early pregnancy and we will explore whether this is a "problem" that needs attention or even medical intervention. Particular attention will be paid to the role of body fat (as proxy for positive energy balance) in onset of menarche.

In Chapter 2 I will discuss the evolutionary novelty of highly frequent menstrual cycling, noting that the estimated number of menstrual cycles for women in the ancestral past is close to 100, fewer than a quarter the number many women in health-rich nations experience. What are the health consequences of 12–13 rises and falls of reproductive hormones and the associated cell turnover rates per year for more than 30 years? Chapter 3 focuses on conception and early pregnancy noting that there may be selective advantages to early pregnancy loss when embryos are not likely to survive pregnancy and childhood or when conditions for the mother are so dire that successfully raising a child may be almost impossible. In some cases, an early miscarriage, sad as it may be for many, may be a "good thing." Other complications of pregnancy such as gestational diabetes and preeclampsia are also given an evolutionary medicine twist.

Getting pregnant is one thing, but staying pregnant is yet another story, and evolutionary medicine has a lot to say about the later stages of pregnancy. In Chapter 4 I will discuss morning sickness, stresses on the bipedal body caused by carrying a baby in the last few months of pregnancy, and factors that interfere with fetal growth and that may have lifelong consequences. The challenges of giving birth to a large-brained infant through a small bipedal pelvis are the focus of Chapter 5. Other medical consequences of bipedalism, especially ones that impact women, are also discussed in this chapter. Although cesarean section has saved the lives of millions of mothers and babies, I will conclude this chapter by considering the consequences of elective c-sections for both mothers and babies.

Chapter 6 considers the extreme dependency of the newborn infant and the all-important things that happen in the first hour after birth that influence infant development, maternal health, and mother-infant bonding. The unfortunate phenomenon of baby blues and its more extreme form, postpartum depression, are also discussed, even though these do not manifest until several days or weeks after birth. Chapter 7 has the provocative title "women are defined by their breasts" so it is obvious that this chapter will focus on breastfeeding and its consequences for both infant and maternal health. Chapter 8 continues this discussion and also provides a critique of the medical concept of "normal" infant growth and the appropriateness of sleeping in the same bed with an infant.

One of the major themes of this book is that women's bodies are shaped for enhanced reproductive success, so trying to understand the early termination of reproductive function at menopause is a challenge. This is the topic of Chapter 9 in which I review various theories for why menopause occurs, whether it is unique to humans, and what selective pressures keep women alive and healthy when they are no longer able to reproduce. The experience of menopause is another aspect of women's biology that is heavily influenced by cultural values and expectations. Once a woman has ceased reproducing, is she of any value in an evolutionary sense? Of course the answer is "yes" and I will provide support for this positive response in Chapter 10, where I focus on the very important roles that grandmothers play in increasing reproductive fitness for those who share their genes. I will also discuss factors that influence longevity.

Finally, in Chapter 11, I try to summarize the book with an emphasis on what knowing about the evolution of women's reproductive bodies can tell us about women's health in the 21st century. A concept that runs through many of the chapters is that the hormonal profiles of women from health-rich populations are actually at the extreme end of a range of variation in apparently healthy reproductive functioning. These are the profiles that are generally regarded as "normal" by medical researchers and clinicians, but they may be unusually high when considered in the context of evolutionary history. The consequences of these high levels may account, in part, for other health phenomena like premenstrual syndrome (PMS), postpartum depression, menopause "symptoms," osteoporosis, and reproductive cancers (breast, ovarian, and endometrial) in women and their daughters, prostate cancer in their sons. Rates of these health concerns are common where levels of reproductive hormones are high and I doubt that their association is mere coincidence. Although evolutionary medicine is not in the business of making clinical recommendations, I try in this last chapter to make some suggestions for what we can do to improve women's health based on our understanding of women's evolutionary history.

1

Are We Grown Up Yet?

Most people who are currently in their 70s or 80s have seen amazing changes in technology, wealth, transportation, and health in their lifetimes. For example, in the 1940s very few people routinely flew on airplanes. Only about 44% of Americans owned their own homes in 1940 compared with almost 70% in 2006 (U.S. Census Bureau). Life expectancy in the United States was 63 in 1940 and was almost 78 in 2004.

Another change that demographers, pediatricians, and parents have paid attention to is the age at which girls reach sexual maturity. If you ask women centenarians when they had their first menstrual period, most will give figures that are several months to several years older than the age girls first menstruate today. In the 1940s, girls in the United States started menstruating at about age 13, but today healthy American girls are having their first periods about six months earlier. Another is a parallel (although inverse) increase in stature documented over the same period. The decrease in the age of onset of menarche and the increase in adult stature recorded in health-rich nations appear to be related to recent improvements in diet and health care and are generally assumed to be signs of well-being (Figure 1-1).

But does early menarche really mean good health? I guess it depends on what we mean by "good." Cross-cultural surveys indicate that girls from higher socioeconomic classes tend to mature earlier than girls from poorer classes within the same communities. Because girls from higher socioeconomic classes tend to be better nourished and to have had better health care throughout their lives, the inevitable conclusion is that early menarche is better. Furthermore, early menarche is linked to greater longevity, another indicator of good health.[1] But things may not be that simple, and to examine this further, we need to know what factors trigger menarche, the beginning of ovarian function.

Growing Up

The years between childhood and adulthood are variously called adolescence, puberty, teenage, youth—the terms tend to be used interchangeably in popular

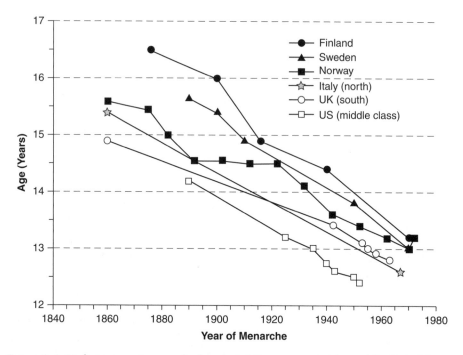

Figure 1-1 Declining age at menarche for selected European countries and the United States. From Trevathan et al., 2008, page 35.

discourse. The term *puberty* is usually invoked to describe physical changes that occur between childhood and adulthood; puberty is fairly well defined for most mammals and simply refers to the onset of reproductive function. In humans, the period following puberty is marked by an accelerated rate of skeletal growth that is referred to as the "adolescent growth spurt." This sudden period of accelerated growth does not seem to happen in monkeys and apes as they approach maturity, and anthropologists Barry Bogin and Holly Smith have suggested that the adolescent growth spurt is unique to humans.[2]

The main thing that is growing during the adolescent growth spurt is the skeleton, especially the bones of the legs and arms. The pelvis is also undergoing its final transformation that results in marked differences between the sexes. In fact, up until puberty, it is extremely difficult to tell the difference between male and female skeletons in an osteological or forensic sample. But after growth is complete, the pelvic bones are probably the very best for distinguishing the sexes. Of course these differences relate to the different "jobs" that the pelvic girdle does, especially in serving as the passage for childbirth in women, as will be discussed extensively in Chapter 5.

Humans, especially females, have a fairly long period of what is often (and erroneously) referred to as "adolescent sterility" (as noted in the introductory chapter with regard to fertility and fecundity, this period is more appropriately called "adolescent subfecundity") following menarche in which most, if not all, menstrual cycles are nonovulatory. This ambiguous period between the appearance of secondary sexual characteristics (enlarged breasts in girls, pubic and underarm hair in both girls and boys) and completion of growth is what has usually been defined as adolescence in humans.[3]

In much the same way as with infancy, adolescence is "used" by cultures to shape the child into the adult that best fits with the cultural ideal.[4] In some cultures, adolescence may be seen as a short period (or not existing at all) or relatively long, depending on how much enculturation is to take place. In health-rich cultures where adolescence may last 10 years (from menarche at age 11 to the age at which alcohol can be legally purchased in the United States and formal education is usually terminated), this long period may be necessary for a person to acquire the social, intellectual, and technological skills necessary to function as an adult. It should be noted that in some cultures, attainment of adult status depends on very specific biological (for example, the birth of the first child) or cultural (circumcision, marriage) events that may be somewhat independent of physical developmental processes.

What does the decreasing age of menarche say about childhood, particularly in girls? As noted earlier, childhood is often defined as that period between weaning and puberty, but given the variation in age of each of these "brackets" how can it be a defined life cycle phase? Is an 11-year-old girl who has reached menarche no longer a child? And does the 15-year-old who has not yet menstruated for the first time remain a child while her peers are called "adolescents"? This debate over what distinguishes adolescence from puberty fits well with the biocultural model that makes the most sense for viewing the human life course. The period between childhood and adulthood is both a biological and a cultural stage and we cannot understand one without knowing about the other.

Trade-Offs in Growing Up

In the introductory chapter, I discussed life-history theory, noting that animals have only so much energy to devote to growing, staying alive, and reproducing. They must budget that valuable resource in ways that ultimately lead to increased reproductive success, the only measure of success as far as evolution is concerned. Beginning at birth, energy is devoted to growing and maintaining health sufficient to reach adulthood. But at some point, growth can begin to take a back seat to reproduction. That is what begins to happen at puberty. In addition to growth, however, human children also spend a lot of time in their growing years learning how to be members of their societies, how to make a living, and how to care for their own children. Given the complex social worlds of humans and the challenges of working and raising young, it takes a long time to learn the skills necessary for successful adult- and parenthood. That is one of the reasons it takes 10–12 years to grow up once a child has been weaned from full nutritional dependence on the mother.

As a child is growing, most of the energy available goes into skeletal growth. So if illness is rare and diet is sufficient, "normal" growth will occur. Weight gain occurs only if there is more than enough energy for growing taller. In the past, energy intake was such that there was not a lot left over for putting on fat until skeletal growth had been completed. Unfortunately, in much of the world today, caloric intake is often excessive, meaning that children are not only growing in height but are also putting on weight (as fat) at the same time. This has led to what has been described as an epidemic of childhood obesity. For example, according to the National Health and Nutrition Examination Survey (NHANES), in the United States in 2004, almost 20% of children and teens were overweight, and it appears that the trend toward overweight and obesity is increasing.

Genes also contribute to growth, although they seem to be secondary to environmental and dietary influences. Short parents tend to produce children who are also short when they reach adulthood, and genes probably put an upper limit on the height a person can achieve. But if a child has "tall genes" and experiences poor health and nutrition in childhood, she will be shorter than her genetic potential. And if the parents are short because their growing-up environments were impoverished, they often have surprisingly tall children if their life conditions improve. This was illustrated well by Guatemalan Mayans who immigrated to the United States at the end of the 20th century. Anthropologist Barry Bogin and his colleagues report that the children of these migrants who grew up in the United States are taller, heavier, fatter, and more muscular than children from the same population who remained in Guatemala.[5] Despite an illustrious past history, the Mayan people of Central America have experienced generations of poverty and oppression by the dominant Latino (which Bogin refers to as *Ladino*) cultures. Those that migrated to the United States because of political unrest in Guatemala have maintained many of their cultural traditions and language, but their children go to school in the United States and grow up in environments of relative prosperity[6] (although many are still impoverished by U.S. standards).

In their homeland the Maya are typically short and have even been described as American "pygmies."[7] The term *pygmy* usually implies a genetic limitation on growth, but Bogin's studies have demonstrated that it is not genetic; rather, economic and political factors can explain the Maya's small stature. This illustrates one of our favorite refrains about humans, and that is their plasticity. Similar changes in stature have been reported for Africans subjected to apartheid in South Africa and for migrants to the United States, Canada, and the United Kingdom from southern Europe, Mexico, and South Asia where standards of living have improved in one or two generations. In many cases, the trend is so pronounced that in a few generations, the stature of the descendents of the formerly short migrants is the same as the host population unless economic oppression greeted them in their new homes. In countries that have experienced rapid industrialization since World War II, young adults today are several inches taller than their grandparents.[8]

Most ecologists discuss adverse effects on growth and development in terms of acquisition of resources and avoidance of illness or predation, as discussed earlier. But human growth can be affected by a number of other factors, most of which are the result of cultural impacts. Children who grow up in the midst of war, racism, and oppression, or who live at high altitude, will also experience slower growth and shorter stature.[9] Those who work hard and carry large burdens also show signs of slowed growth. Girls are often treated more harshly than boys, with the result that adult size differences between men and women are greater than would be expected based on genes alone. Finally, migration often indicates conscious decisions by parents to improve the lot of their children and, in the language of evolution, improve their reproductive success.[10] It is not clear whether nonhuman animals make such decisions about parental investment.

Other Changes at Puberty

I have mentioned skeletal growth that occurs at puberty, but a number of other changes also take place that mark the entry into the reproductive phase of life. Pubic and underarm hair begins to appear in both sexes, breast development begins in girls, and facial hair appears in boys. Boys find it easier to develop their muscles and girls find it easy (sometimes too easy) to accumulate body fat. Boys start sounding like men. All of these changes are under hormonal control, so what underlies them is the maturation of the ovaries and testes.

This maturation phase is happening to both boys and girls, but it tends to occur 1–2 years earlier in girls, a fact that many of us remember from our junior high school years when social dancing during physical education classes was a bit awkward. And within the sexes the changes occur at different ages, which often put social and emotional stress on the late maturers, especially boys. The phenomenon of the two sexes maturing at different rates is called *sexual bimaturism* and is seen in other primates as well. Anthropologist Carol Worthman points out, however, that when considering all aspects of physical and physiological maturation, boys are only a few months behind girls—it is just that the ways in which girls are changing are much more obvious.[11]

Though it is amazing how much we do know about human biology, we still do not fully understand what leads to maturation of the reproductive system or what causes the labor of childbirth to begin. In other words, hormonal characteristics of humans are still uncharted territory in many ways. Reasons for this are somewhat obvious because a full understanding would require experimentation and testing of competing hypotheses by using living humans, which would not be acceptable. Although there are many similarities in biology, our reproductive systems are so different from those of mice, monkeys, and even chimpanzees that experimenting with these species, even if it were not so controversial in our society, would not necessarily produce the information we need.

Reproductive maturation is a two-stage process: *adrenarche*, when the adrenal glands mature at about ages 6–8 (earlier for girls) and *gonadarche* (menarche in girls), when the gonadal hormones kick in. Formerly it was believed that the two were linked and that timing of adrenarche could predict timing of puberty, but it is now believed that they are not coupled. We will see that gonadarche is actually the reactivation of a system that is busy during fetal development but remains inactive for a decade or more following birth. At puberty, gonadotropin-releasing hormone (GnRH) begins secreting, stimulating the release of luteinizing hormone (LH) and follicle-stimulating hormone (FSH), which initiate maturation of the other systems related to reproductive function in adulthood (such as breast development, underarm hair growth, pelvic shape and size changes). Menarche actually comes somewhat late in the cluster of events that occur at this time.

What Does Fat Have to Do with It?

Most women who look back over their experiences in junior high and high school can recall that the girls who were a bit plumper and less active than the thinner and more athletic girls had their first periods earlier, often by several years. Thus, the suggestion that accumulation of body fat had something to do with menarche made sense when it was first proposed by population scientists Rose Frisch and Roger Revelle in 1970.[12] They examined data for growth in girls and concluded that achieving a specific weight triggered a cascade of physiological events that resulted in the onset of puberty and menstruation. It was common knowledge that when women experienced rapid weight loss or engaged in strenuous activities they would often skip menstrual periods, a phenomenon known as *amenorrhea*. Amenorrhea is also associated with anorexia nervosa in girls. When the girls gain weight again or stop strenuous exercise, their periods usually resume. Later, Frisch and endocrinologist Janet McArthur proposed a minimal weight for resuming menstruation in girls for whom it had stopped.[13]

There was a great deal of criticism of what became known as the "Frisch hypothesis," but the general argument that body fat accumulation influences reproductive potential in some way makes evolutionary sense, as well as common sense, if we think of body fat as proxy for positive energy balance. Consider again the arguments from life-history theory. There is only so much energy to go around and, at least in the past, most of it was used for growth in the first two decades of life; and only when growth was completed was there sufficient caloric reserve to direct toward reproduction. In other words, a girl's body could not "afford" to both grow herself and grow a baby, or both processes would be compromised, as would her lifetime reproductive success. So it makes sense that when calories start being stored as fat, this is a signal that they are no longer needed for growth and it is time to start reproducing. And once reproductive maturity is reached, if illness or starvation occurs so that energy reserves are needed for survival, the reproductive system shuts down to avoid competition for calories. When the crisis passes and extra calories are again stored as fat, the signal is given to resume ovulation. So it may not be that a particular level of fat is the trigger, but it is safe to assume that fat plays some sort of role, at least as an indicator of energy availability.

Not only does the relationship between fat and fertility seem to be "common sense" to us today, but it may have also been obvious to our ancestors if the so-called fertility figurines are any indication. Frequently found in the archaeological record of the Upper Paleolithic are small clay or stone figures such as the Venuses of Willendorf, Laussel, and Dolní Věstonice that have been interpreted to relate to fertility. Of course, if the interpreters think that fat and fertility are related, then it is not surprising that they would see the figurines as fertility amulets, so the argument is circular. But in any event, the figures show exaggerated features (breasts, hips, and thighs) that may indicate a special regard for women with excess body fat.

It would be hard to overemphasize the importance of energy reserves and related food availability for successful reproduction. But fat or extra energy availability is not enough. There are, sadly, many girls who at ages six or seven weigh as much and have as much body fat as girls in their late teens, but they are not menstruating nor are they capable of having babies. So for these girls, the body is not sending a signal that there are plenty of calories for pregnancy. One reason that they are not getting pregnant (with rare, probably pathological, exceptions) is that it would be impossible for them to give birth because their pelvises are not fully formed. There must be some other signal that the body needs to begin reproductive functioning and that signal seems to be completion of skeletal growth. Peter Ellison[14] has proposed the "pelvic size" hypothesis as the trigger for reproductive maturation. This makes even better evolutionary sense than the fat hypothesis because without cesarean section, girls whose pelvises had not matured would not have been able to deliver babies even if they had gotten pregnant, so natural selection probably acted to decrease the likelihood of onset of sexual and reproductive maturity preceding skeletal maturity. The best scenario from an evolutionary perspective then is this: first use all the energy you have to grow the skeleton and the pelvis so that it will be possible to physically deliver a baby if you get pregnant. Once skeletal growth is complete, take any leftover energy and store it as fat so that it can be used to maintain a pregnancy once that is achieved. Then, all it takes is a successful ovulation and sexual intercourse and, voila! There is a chance for reproductive success.

On the other hand, the idea that excess energy reserve (or positive energy balance) is necessary for reproduction flies in the face of evidence to the contrary for many species. How can we forget the photos of starving women in places like Darfur or Zimbabwe who are nursing their infants? Clearly these women were not likely to have accumulated excess body fat a year or so earlier so they must have ovulated and successfully carried a pregnancy to term despite low levels of food intake. And in fact, examination of the lives of most mammals in the wild suggests that reproduction continues despite sometimes severe energy stress.[15] Starvation is not something that we are directly familiar with in the West (except in the tragic cases of anorexia nervosa), but we are familiar with teenage girls putting on extra weight and then starting their periods, so it is not surprising that we think that fat and reproduction are related.

What seems to be going on is that food intake has an almost immediate effect on GnRH secretion, which in turn influences the release of LH and ovulation.[16] So a few extra calories beyond what are needed by the body to stay alive are often sufficient to stimulate ovulation, even if there is no way of actually seeing (in weight gain) the few extra calories. Thus, it is reasonable to conclude that extra calories beyond what are necessary to live are used for normal reproductive functioning, but storage of those extra calories as fat is not required. In fact, it can be argued that for most of human evolutionary history there were few times in which food availability would lead to fat storage, yet clearly our ancestors were able to reproduce.[17] On the other hand, fat is a clear signal that there are extra calories, but that luxury is likely quite recent in human history, "Venus" figurines notwithstanding. If natural selection had required lots of body fat to ovulate, most of us would not be here today. But our ancestors who could readily store excess calories as fat would have been more likely to survive and successfully reproduce during the lean times, so that ability is with us today, often with negative consequences for health since the lean times are few and far between for many of us.

As noted earlier, many animals have reproductive systems that are sensitive to seasonal variation in food resources. We do not typically think of humans as among these species because most of us do not experience any real variation in resource availability on a seasonal basis. We also do not usually think about birth seasonality in humans, except perhaps such anomalies as the peak in births nine months after a holiday or an extensive power failure. There is also evidence that in the northern hemisphere, there is a peak in conception rates in the spring when our "hearts turn to thoughts of love." How about women in the evolutionary past, however? Would their physiology have been more in tune with resource availability? As is often the case, since we cannot reconstruct fossil physiologies, we turn to evidence from populations that are living in environments similar to that of our ancestors. One such example is a population of subsistence farmers in The Gambia who seem to show distinct troughs in conceptions at the time that food stores run out and women are working particularly hard to put in next year's crops. At this time, their daily caloric intake is around 1,500. The peak of conceptions seems to correlate with accumulation of body fat sufficient to support pregnancy and lactation or, more likely, a period of even slight movement into positive energy balance.[18]

Fat on the Hips and Buttocks

An interesting proposal that still needs further testing is one by anthropologists William Lassek and Steve Gaulin, who argue that it is not the amount of body fat that is important but the way in which it is distributed on the body. They propose that menarche is associated with increasing amount of fat in the hips and buttocks and that fat deposition in these areas is a better predictor than skeletal growth or total body fat.[19] Hip and buttocks fat appears to produce more leptin, a hormone that influences both ovarian function and skeletal growth. Furthermore, the amount of leptin is inversely related to age of menarche, suggesting that fat accumulation on the hips and buttocks may be the trigger for menarche.

Lassek and Gaulin also argue that hip and buttocks fat are the primary sources of fatty acids that are passed from the mother to the fetus during gestation and the infant during lactation. These long-chain polyunsaturated fatty acids (LCPUFAs) play a major role in brain development. Thus, they argue, hip and buttocks body fat does not just signal overall energy availability for pregnancy but also signals that the essential fatty acids for brain development are in sufficient supply. Does hip and buttock fat correlate with cognitive abilities? In an examination of the third NHANES study database, Lassek and Gaulin found that high amounts of hip and buttocks fat relative to waist size (a low waist-hip ratio) was predictive of women's own and their offspring's cognitive performance.[20] They refer to this fat as "a privileged store of neurodevelopmental resources."[21] Clearly there is a lot more than hip and buttocks fat that contributes to cognitive abilities, but it is intriguing to think that this specifically female characteristic may be among the factors.

One concern with teen pregnancies is that the pregnant teen and her fetus are competing for the same resources, with one result being lower cognitive performance scores for their children in comparison to the children of women who had completed their own growth before becoming pregnant. Lassek and Gaulin found in the NHANES III database that teens with large hips and buttocks and small waists showed a protective effect in that their children had cognitive scores similar to those of nonteen mothers.

Finally, Lassek and Gaulin propose that this is the reason that men in many cultures judge as attractive women who have high amounts of fat on their hips relative to their waist size (low waist-hip ratio). For example, anthropologists Peter Brown and Mel Konner in a cross-cultural survey of 58 cultures found that men in 90% of them found fatter legs and hips to be attractive.[22] One conclusion is that the sexual dimorphism seen in body fat has evolved not just because men found it attractive in women but because fat on the hips produced more LCPUFAs that could be directed to fetal brain development. Fortunately, men found that attractive, and their children were not only smarter, but their daughters were more likely to grow up to have fat distribution best for attracting mates and producing smart babies. So if you have more fat on your hips and buttocks than you like, you have got your ancestors to blame—or thank.

Psychosocial Factors Affecting the Timing of Menarche

In addition to overall health and nutrition, a number of demographic, social, and emotional factors seem to have an effect on the timing of menarche.[23] For example, girls in rural areas tend to reach menarche later than those in urban areas. Family size and age of menarche seem to be directly related, with girls in larger families menstruating later. And, as noted, socioeconomic factors have an influence, most likely through diet, activity levels, and health care. Psychosocial stress in general, often resulting from socioeconomic and family challenges, also has an impact on reproductive functioning, with the primary effect of lowering age of menarche and age of first reproduction.[24]

In recent years there has been a lot of interest in the impact of social, psychological, and familial circumstances on timing of menarche. The mammalian literature is replete with evidence that exposure to adult females can delay puberty and exposure to adult males can accelerate it. The proposed mechanism is pheromonal.[25] One view of the effects of pheromones is that changing social circumstances that lead to more women working outside the home can explain the downward trend in age of menarche in the past few decades. The idea is that, similar to what has been described for other mammals, when women of reproductive age are in daily and frequent contact with prepubescent girls, their presence (their pheromones) delays onset of ovarian function in the girls, but now that most mothers work outside the home, their pheromones are not having the same effect.[26]

Anthropologists Pat Draper and Henry Harpending, in a review of the cross-cultural literature, propose that girls who experience different types of family structures, primarily father present or father absent, have different developmental trajectories.[27] They suggest that girls who perceive males as unreliable for child rearing will embark on a reproductive strategy that leads to early sexual activity and childbearing, whereas girls who perceive fathers as present and reliable will delay childbearing until they secure an equivalently reliable mate. The behavior of early sexual activity is associated with early menarche and may be a physiological response to father absence.[28] Psychologist Michele Surbey offered support of this hypothesis in a study of more than 1,200 girls, among whom those who experienced father absence matured earlier than those who were raised by both parents. Furthermore, the younger the girl was when her father left, the earlier her menarche, and if she subsequently lived with a stepfather, menarche was even earlier. In other words, long-term exposure to related males seems to delay menarche, whereas exposure to unrelated males appears to accelerate it. If indeed early menarche is associated with an increased risk of breast cancer, it could be argued that being raised by two parents is protective for a number of health risks. This may be a stretch, however, because the both-parents-raising-children model that we have in the United States is far from universal across cultures.

Psychologist Jay Belsky and his colleagues offer an evolutionary take on the father-absence hypothesis and suggest that it is associated with a reproductive strategy that emphasizes quantity over quality. Recall that when environments are stable, producing low numbers of offspring with high survival rates (going for quality) is adaptive, but when environments are unstable, the best option would be to go for quantity and hope some survive. Households with father absence, they argue, are generally less stable than those with father presence.[29] They highlight stress as the most likely mechanism. In evaluating this proposal, it is important to consider the decade in which it was made, a time when divorce rates were rising and concern about single mothers raising children was great. Today, when more than half of all children grow up in single-parent or stepparent homes the proposal may not resonate as well. Of course, the other side of that coin is very low ages of menarche for almost all girls. These ideas about the effects of family composition on reproduction in girls are controversial and remain at the exploratory or hypothesis-testing stage of thinking, but given the wide diversity of family types seen in contemporary populations, continued testing of competing ideas about menarche is likely.

What Is Good about Early Menarche?

It seems reasonable that the more years of reproductive capability a female has, the greater will be her reproductive success. Thus, one would expect that women with earlier menarche would have more offspring. In support of this, anthropologist Monique Borgerhoff Mulder found that among the Kipsigis of Kenya, women who reached menarche between the ages of 12 and 14 years had almost three more children than those who were older than 16 at menarche.[30] Part of the effect was due to higher offspring survival for the earlier maturers. Related to this is that the early maturing women, as predicted from other studies, came from wealthier households with better access to good nutrition. Another factor that likely plays a role is that earlier menarche is associated with fewer years of adolescent subfecundity,[31] which contributes to higher fertility.

The timing of resource availability may influence reproductive maturation. For example, girls who grow up in impoverished environments and then experience improved diet and health care often start to menstruate soon after their circumstances change and earlier than their peers who remained in the impoverished environment. Anthropologist Carol Worthman suggests that this is evidence of not only the flexibility and sensitivity of the reproductive system but also its ability to respond quickly when conditions for reproducing are significantly improved.[32] These effects of migration will be discussed further in the next chapter.

What Is Bad about Early Menarche?

Although early menarche is related to positive health indicators, there are a number of negative aspects to reaching puberty early. For example, spontaneous abortions were 1.5–2 times higher among Norwegian women aged 20–45 who reached puberty before age 12 compared with those who were over 14.[33] In another study, American women who reached menarche before 12 or after 13 had more unsuccessful pregnancy outcomes than those who were 12–13.[34] Other studies have shown that early maturers have more problems with severe menstrual cramping and more frequent irregular cycles.[35] Preterm deliveries also appear to be more common among early maturers.[36]

Other health consequences linked to early menarche include greater risks for later-life obesity, a variety of reproductive cancers, depression and anxiety, and substance abuse.[37] As noted earlier, early maturers begin regular ovulatory cycles sooner than later maturers, and they have higher levels of some hormones, including FSH and estradiol, than later maturers at the same number of months past menarche.[38] As we will see, this may be the reason that girls who reach puberty early are at higher risk for breast cancer: they not only have more years of exposure to estradiol but their absolute levels are higher once they start regular ovulation.

As noted earlier, diet is related to puberty and one interesting study indicates that the amount of animal protein and fat consumed in childhood impacts the age of menarche, with higher animal-based diets being associated with younger ages of menarche.[39] Increased milk consumption, especially in populations where milk has not traditionally been part of the diet, also has an effect on menarche and later-life health.[40] This may be one of the pathways through which younger age at menarche puts women at higher risks of chronic late-onset diseases and disorders such as obesity and type 2 diabetes. The links among low dietary fiber, high animal protein and fat, early menarche, and breast cancer can be understood in the same way.[41]

Women who reach puberty early and have high levels of ovarian hormones during pubertal development and pregnancy have daughters who are at higher risk for breast cancer when they are older.[42] When a female fetus is developing she is exposed to the circulating hormones of her mother, and this influences the development of her own breast buds in early gestation. The same thing seems to happen with regard to ovarian cancer in that female fetuses exposed to high levels of hormones while their ovaries are developing are also at higher risk.[43] The predictors of this phenomenon are broad hips that develop at puberty under the influence of ovarian hormones, but primarily in women who have early menarche and who are short. Short stature and early menarche occur when girls are undernourished for the first few years of life but then become well nourished as they approach puberty, as often happens with migration or adoption. Thus, as with several aspects of reproductive functioning that I will discuss, maturation events and the sociocultural impacts on them have transgenerational effects.[44]

Early menarche may also have negative consequences for mental health in some adolescent populations. Concern about mental health issues in adolescence has led to a number of studies assessing the effects of early puberty. Although not all of the studies have found support for the proposed relationship, there has been sufficient evidence to alert mental health professionals to be on the lookout for issues that might arise when they are working with young women who achieve sexual maturity at an early age. For example, a U.S.-based study of more than 1,700 high school students found that early maturing girls had higher rates of self-identified depression, anxiety, behavior disorders, and substance abuse than their later maturing peers.[45] Perhaps the most alarming finding has been the relationship between early maturity and suicide attempts in girls in the United States, Finland, and Norway.[46] In a carefully controlled study of girls in North Carolina (USA) that depended on objective measures (rather than self-report) of puberty and depression, the relationship between early menarche and depression was not supported, although puberty itself was associated with depression in girls.[47]

Another concern about early onset of puberty is that in many ways there is discordance between the physiological age of, say, a 15-year-old girl who is sexually mature and what her parents, physicians, and community expect from her. To them, she may be still a child, but some of her behaviors and her reproductive potential are those of an adult. In Carol Worthman's words, the "schedules of childhood" in the minds and texts of an older generation do not fit with what the young adolescents are experiencing,[48] and behavioral conflicts can result, compromising family and individual mental health.

When Should Puberty Occur?

In summary, here is just a partial list of what menarche may depend on: when a girl's mother reached puberty; how much body fat she has; her waist-to-hip ratio; how much fat is in her diet; how many siblings she has; whether she lives with her father or other older men; what she weighed at birth; how physically active she is; whether she has gained weight recently; whether she lives in a rural or urban area; how much psychosocial stress she has experienced; her socioeconomic status; long-term illnesses that she has had; and dozens of things we have not even thought about. In other words, it depends on how a specific genotype and phenotype develop in the specific environment in which one lives. Since the various combinations of bodies and environments are virtually infinite, it is impossible to provide a definitive answer to the question of when is the best time for menarche to occur, given the various costs and benefits associated with age of onset. We can probably talk about the ages at which any concept of "normal" is exceeded (say, less than age 8 or 9, later than age 18), but even those ages may be normal under certain circumstances (like severe malnutrition or over-nutrition). Evolutionary biologist Stephen Jay Gould often said that evolution was a "contingent process." This is no less true for development. So what is the best age for girls to reach menarche? The answer is, as always, "it depends."[49]

Because of the concern about early menarche putting a girl at risk for reproductive cancers, psychosocial disorders, and early childbearing, there has been some discussion of developing interventions that would artificially delay puberty.[50] Unfortunately, as we learned with hormone replacement therapy (HRT) for postmenopausal women, manipulation of "natural" hormonal milieus may have far-reaching, sometimes negative, results. But given the myriad of negative consequences that follow upon childhood obesity, finding ways of keeping young girls from putting on excess body fat may delay menarche and also have positive effects on other aspects of later-life health, so it is hard to argue against interventions to reduce obesity. In fact, as obesity rates in various parts of the world increase, we are likely to see more and more girls reaching puberty at young ages and perhaps higher rates of reproductive cancers in the future. There is also probably not a negative side to making every effort to reduce psychosocial stress in the lives of children and adolescents. And finally, as with almost every other social or health problem we face today, reducing the number of families living in poverty would go a long way toward more healthy ways for maturation to unfold.

To Grow or Reproduce? Teen Pregnancy and Motherhood

One question for a maturing female that derives from life-history theory is whether to grow or breed. In other words, if energy is limited, is it better to use all that is available for growth and maturation or to allocate some to reproducing? Is it better to grow to a larger size before reproducing or to have babies while you are still small? Again, like everything else in life-history theory, the short answer is that it depends on the circumstances. In most instances, larger parents produce healthier offspring, so if the environment is good and somewhat predictable, it might be best to delay reproduction until growth is completed. Furthermore, offspring mortality decreases with mother's age of maturity, so if conditions are favorable, the best strategy would be to delay maturity and reproduction until the point that offspring survival is greatest.[51] But as we well know, conditions for reproducing are not always favorable, so natural selection has favored a range of "options" for growing and reproducing.

Later I will discuss pregnancy extensively, but I have chosen to address teen pregnancy here because of its close association with early onset of menarche and sexual activity. Teen pregnancy is often cited as a major social problem in the United States, with justification for trying to do something about it provided by the evidence that health outcomes are compromised by too-early childbearing. In one view of the world, delaying pregnancy until late teens or early twenties, after education and livelihood have been secured and a suitable marriage partner is found, is optimal for child health, development, and cultural achievements. Not surprisingly, this view is held by people who are in the middle and upper echelons of society and who hold political and economic power. These are the people who expend great effort and resources in well-meaning (usually) attempts to do something about the "problem" of teen pregnancy.

That having been said, there are several reasons that giving birth before a girl has finished growing herself is problematic. For example, the rate of obstetric complications resulting from incomplete growth of the pelvis is much higher in young teen mothers (ages 10–15) than in older women. The birth canal grows at a different rate from the skeleton as indicated by stature. Specifically, when a girl reaches her full stature, her pelvis seems to lag several months behind and will not reach adult size and shape until several years after maximal stature is reached.[52] If girls get pregnant before the pelvis has finished growing, they tend to have smaller babies, but the babies' head size is apparently not smaller, so with smaller birth canals, cephalopelvic disproportion (CPD) is common, resulting in a higher incidence of cesarean section. Furthermore, in one sample that was studied, girls who matured later had significantly larger birth canals and greater stature than the girls who reached menarche early.

Given that there is a limited amount of energy that can be devoted to growing and reproducing, in an ideal world, reproducing would follow upon completion of growth because otherwise, the pregnant girl and her fetus are competing for the same resources with the result that both often come up short. As evidence, babies born to teens tend to weigh less than those born to older women and a higher percentage of low birth weight babies are born to teens,[53] although this may be more due to lack of prenatal care and socioeconomic circumstances than to maternal age. But even when nutrient availability is high, as it is in well-fed American teens, competition between fetus and mother may result in low infant birth weight. In sheep, overfeeding of young pregnant dams while they are still growing results in smaller placentas and lambs, which could explain the process whereby fetal growth is limited in humans who are rapidly growing in food-rich environments.[54] Anthropologist Karen Kramer suggests that "humans are metabolically designed to keep growing at the expense of a fetus, even when energetic constraints are relaxed."[55]

The magnitude of the risks of teen pregnancy is hotly debated, with some attributing negative outcomes almost exclusively to socioeconomic status and others attributing it to incomplete maternal growth.[56] One thing that could be accounting for contradictory findings with regard to birth outcomes is the tendency in national statistical surveys to lump all teens in one category (ages 13–19), when it is obvious that girls in the lower ages (13–14) and upper ages (18–19) are two very different organisms. In fact, obstetric outcomes for girls 18 and 19 are about the same as for women in their 20s. In one study, the authors attempted to define early teen childbearing by considering the point at which rates of infant mortality, very low birth weight (less than 1,500 grams), and very early delivery (earlier than 32 weeks) begin to stabilize in the United States and identified that point statistically as age 16. Thus, they suggest that, in general, the ages for concern about early childbearing should be those who give birth at 15 years or younger.[57]

Among the Pumé, a foraging population in South America, girls who become pregnant when they are younger than 14 are four times more likely to lose their infants than girls who are older than 17 at first pregnancy.[58] This is a population where birth control is not practiced, sexual activity begins soon after menarche, socioeconomic status does not vary, prenatal care is not provided to any pregnant women, and early childbearing is valued and encouraged. Thus, most of the confounding variables of concern in health-rich nations are not present among these people, but birth outcomes for the very young teens are still compromised. It should be noted, however, that these girls reach puberty at younger ages than might be predicted by the good health and nutrition models, lending support to the view that early menarche can be triggered by various forms of stress.[59] As noted earlier, this is the population in which Kramer found that the women who began reproducing in their mid-teens (14–16) had the greatest lifetime reproductive success.

From an evolutionary perspective, delaying pregnancy until the "best" time makes sense when resources are abundant and environments are stable. But for people whose lives are characterized by poverty, ill health, and powerlessness, reproducing early may be the most reasonable strategy, and almost any time after menarche may be "good enough" for having children. Among African Americans living in poverty, for example, rates of low birth weights and infant mortality are lowest for mid to late teens and rise throughout the 20s and 30s[60] as grinding poverty and community violence continue to take their tolls on women's health. If a young woman doubts that she will be alive when her children are grown, she may hedge her bets and have them as young as she can. As a teen, she may also be in better health and able to provide better care for them than later, when her health declines. Furthermore, if educational and employment opportunities are rare or absent and stable partners are uncommon, delaying childbearing may not be a bad thing.[61] One study found that the availability of a marriage partner was a better predictor of early pregnancy than whether the woman was European American or African American,[62] supporting the idea that environmental stability plays a more important role than other aspects of sociocultural background.

Behavioral scientist Arline Geronimus, one of the foremost spokespersons for the view that it is largely dominant social classes that define teen childbearing as a problem, argues that it is not teen pregnancy that "causes" the problems of low birth weight and high infant mortality, but that socioeconomic circumstances predispose both teen pregnancy and poor obstetric outcome. In other words, infant outcomes are compromised among the poor for all ages and maternal age or developmental state (except for the very young) has little to do with it.[63] So efforts to reduce teen pregnancy per se may have little to no effect on infant outcomes. But if resources could be focused on improving prenatal care and resource access for all, birth weights would rise and infant mortality would decrease for all women living in poverty. Unfortunately, public focus will continue to be on preventing teen pregnancy because it "just makes sense," despite evidence that it is not the cause of the problems it hopes to correct.[64] As anthropologist Brigitte Jordan has noted, "The power of authoritative knowledge is not that it is correct but that it counts."[65]

Giving birth when you are young, immature, and financially dependent on others may not be a problem where there are strong support networks and multiple caretakers to help.[66] Where it is the cultural ideal for two adults to marry and raise children as a parental team, a young, single mother may seem like a problem, but where child rearing is done by multiple people across several generations, it may not matter how old the mother is or whether or not she is married.

The length of time between the onset of reproductive function and the first pregnancy is one of the most significant risk factors for breast cancer.[67] In health-rich populations where contraception is the norm, 10–15 years may intervene between these two events, which may partially account for the higher cancer risks in the same population. Perhaps for this particular health outcome, having a baby in your teens is protective. But our ancestors who had their first babies soon after they reached puberty were in their late teens. A "Paleolithic prescription" to prevent breast cancer today by having a baby soon after menarche would mean becoming a mother at age 14 or 15. Maybe there are better ways to minimize breast cancer risks.

As is commonly known, early childbearing often repeats itself in subsequent generations in that young women who begin reproducing in their teens often have grandchildren before they reach the age at which some (usually more affluent) women are having their first babies, in their early to mid-thirties. As noted, early childbearing is often associated with unpredictable life circumstances: unequal access to resources and medical care is a major contributor to psychosocial stresses and life course instabilities. But given its link with low birth weight, early child-bearing also sets a trajectory toward poor childhood and adult health.[68] Medical intervention strategies tend to focus on specific illnesses and health concerns for which there are significant disparities (low birth weight, adolescent pregnancy, diabetes, cardiovascular diseases, hypertension), but success at reducing these disparities will be limited as long as the underlying causes, such as socioeconomic disparities, are not fully considered. A public health paradigm (what behavioral scientists David Coall and Jim Chisholm call "evolutionary public health") that targets reduction in social inequalities will likely have more permanent and long-lasting impacts on the health of children, adults, and future generations.[69] Chisholm and Coall argue that if we expend all of our public health efforts just on keeping people alive without paying attention to quality of life, we will continue to see young people in desperate conditions trying to make the best out of a bad hand by reproducing early.[70]

In the end, it appears that the number of factors that influence the maturation of the reproductive system and first pregnancy in girls is extensive and most seem to be interrelated in ways that are difficult to separate out. This should not be sur-prising, given the biocultural animals that we are. Why would we expect that any single cause (genes, diet, stress) works alone in triggering such a critical life-history event? But given a concern about the almost universally negative consequences of very early menarche, the search for modifiable factors continues.

A major point made in this chapter is that the age of maturation has gradually declined in the last century due to positive health and lifestyle changes, but it is important to recall that there is within every population, a great range of variation. Evolutionary medicine and life-history theory remind us that growing up and maturing are sensitive to local environmental situations (like diet, health care, and parental care practices) and that a textbook-based concept of "normal" adolescence will work only for a small range of the variation observed and primarily what is experienced in the health-rich world.[71] An evolutionary perspective argues that public health measures must look beyond the Western concept of "normal" maturation and "normal" adolescence toward consideration of sociopolitical and socio-economic contexts of development and intervene when necessary to bring about widespread improvements in health.[72] As Worthman notes, "The close associations among resource allocation, pubertal timing, adult competence, and human health make equity in access to resources for child health and development a priority for preventive medicine."[73] In fact, I and others would argue that improvements in health at all levels requires more attention to social context than to medical research.

The next chapter discusses the years of menstrual cycling that follow upon menarche. If I were writing this book from the perspective of ancestral women, the discussion would probably be somewhat meaningless, because menstrual cycling was not something that occupied most of our female ancestors' reproductive years. In fact, for most of the decades between menarche and menopause women were most likely pregnant or breastfeeding, and very few months of their lives were characterized by menstruation. But, as we will see, that one difference may account for a number of health challenges we face in the 21st century.

2

Vicious Cycles

As I mentioned in the Introduction, in evolutionary medicine we often talk about mismatches in which poor health may result from aspects of our contemporary lives that are out of step with our evolved biology. One of the best examples can be seen in the contemporary diets that are often mismatched with evolved nutritional needs. I am particularly concerned about ways in which contemporary reproductive behavior and biology may be mismatched with the reproductive lives of our ancestors, and that may result in negative consequences for health, particularly for women. One familiar example is the dozens of menstrual cycles women in health-rich populations have each year, whereas before the use of birth control, women had very few menstrual cycles. Because most of their reproductive years were spent pregnant or breastfeeding, ancestral women may have had only 100–150 menstrual cycles in their lives.[1] Consider that today, a woman who uses birth control has two or fewer pregnancies, and breastfeeds her infant only a few months or not at all, may have as many as 350–400 cycles between menarche and menopause. Women's reproductive physiology may not be well adapted to the routine monthly fluctuations of ovarian hormones that occur throughout a typical menstrual cycle. As reproductive biologist Roger Short observed, "Since natural selection has always operated in the past to maximize reproductive potential, women are physiologically ill-adapted to spend the greater part of their reproductive lives in the non-pregnant state."[2]

Consider a scenario that has been proposed to describe the life of a woman in the time before agriculture. Based on what we know about hunting and gathering populations, she would have experienced menarche at about age 16, would have had about three years of subfecund cycles (most cycles in the first few years following menarche would have been *anovulatory*, or without ovulation), may have conceived her first child about age 19, followed by two to three years of breastfeeding, and another child about four years after the first. In a healthful environment, she may have had as many as six children, with perhaps four of them surviving to reproduce. She may have nursed her last child for more than four years, perhaps never resuming ovulation or menstrual cycling; thus, menopause would have passed with little notice. For some women in the world today the

41

patterns are somewhat similar, with the exception that infant mortality is much higher and in some areas, perhaps only two of six to seven children born survive to reproduce. Even in the few remaining hunting and gathering populations that have been studied recently, infant mortality is much higher than it is in health-rich nations and than it probably was in the past.[3]

For women in health-rich populations, the scenario described earlier is unusual. As I noted in the last chapter, since the early 1800s, there has been a steady decline in the age of menarche, so that today, girls often experience their first menstruations even before they reach their teens. With sexual experimentation beginning at younger and younger ages, these teens either become pregnant early (often before they themselves have finished growing) or they use birth control to delay pregnancies until their late 20s, 30s, or altogether. Breastfeeding, if it is done at all, often lasts less than one year, and the way in which it is practiced does not reliably prevent ovulation, so women typically resume using other methods of birth control soon after a baby is born. Birth intervals may be as low as two years or as long as six to eight years, depending on preferences of women and couples. Although getting pregnant may be too easy for some women, many others struggle with low fertility or infertility, spending thousands of dollars to achieve pregnancy. Women in health-rich nations tend to limit the number of children to two or three and they use various ways to prevent conception until they reach menopause (conventionally defined retrospectively as one year past their last menstrual cycle) at about age 50.

Table 2-1 provides a summary of these two scenarios with a focus on the total number of lifetime ovulations, as described earlier. The number on the bottom line (160 for women in foraging cultures and 450 for contemporary women using contraception) begs the question of the impact of highly frequent menstrual cycling on women's health. Following Roger Short,[4] women's bodies were not "designed" to be exposed to 400 or more monthly rises and falls in hormones, with the associated effects on cell turnover rates, especially in breasts and uterus. Every time a cell

Table 2-1 Comparative reproductive variables for foraging populations and Americans

	Foraging populations	Americans
Age at menarche	16.1	12.5
Age at first birth	19.5	24.0
Years between menarche and first birth	3.4	11.5
Years of lactation per birth	2.9	0.25
Completed family size	5.9	1.8
Total years of lactation (approx.)	17	0.5
Age at menopause	47	50.5
Estimated total lifetime ovulations	160	450

From Eaton et al., 1999.

divides, there is an opportunity for a cancerous mutation to occur, so frequent cell turnover rates provide more opportunities for mutations. Furthermore, the regular, frequent surges in estrogen likely have an impact on women's health, especially, as proposed by several scholars[5,6] on the estrogen-related cancers of the breast, uterus, and ovaries.

Although comparative rates are difficult to obtain, one estimate is that the rate of breast cancer for health-rich nations, where birth control is practiced and childbearing is limited and deferred, is as high as 100 times the rate for women who are not using contraception and are spending the bulk of their reproductive lives pregnant or breastfeeding in patterns that result in lactational amenorrhea[7] (see Figure 2-1). For these women, the hormonal milieu to which they are most commonly exposed is very different from that of contracepting women (Figure 2-2). This figure serves to illustrate the differences in fluctuation of ovarian hormones over a nine-month period for women who are cycling, pregnant, or breastfeeding. Associated with each rise in progesterone and estrogen are cells dividing in the ovaries, breasts, and uterus in preparation for pregnancies that do not occur.

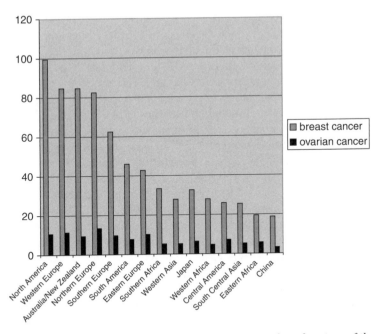

Figure 2-1 Incidence rates for breast and ovarian cancers in selected regions of the world (rates per 100,000; age standardized). Data from Parkin et al., 2005, following Pollard, 2008, page 77.

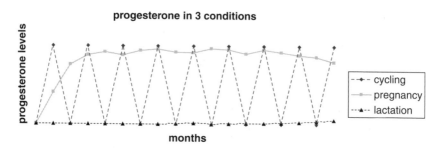

Figure 2-2 Schematic representation of progesterone exposure in nine months of cycling, pregnancy, and lactation. Estrogen exposure would show similar relative values.

Levels of Hormones in the Menstrual Cycle and Pregnancy

Lifetime exposure to reproductive hormones depends not only on how many menstrual cycles a woman has but also on the levels of hormones to which she is exposed in each cycle and during pregnancy. As we have seen, women today experience more frequent fluctuations in ovarian hormones, but, in addition, the absolute levels during each cycle tend to be higher in women from health-rich populations than they are in women from health-poor populations. For example, among the !Kung San of Botswana, average serum estradiol levels in a nonpregnant menstrual cycle are 112 (pg/ml), compared with 136 and 164 for Chinese women from Shanghai and American women from Los Angeles, respectively.[8] A number of studies have found ethnic variation in estradiol levels within the United States.[9]

Progesterone levels are also highly variable across populations with typical values in health-poor populations being about one half to two thirds the levels described as normal in the United States.[10] In fact, the low levels seen in many populations are those associated with women who seek assistance from fertility clinics in the United States.[11] Anthropologist Virginia Vitzthum and her colleagues compared progesterone levels from pregnant and nonpregnant women in the Bolivian highlands with those from women in Chicago. As seen in Figure 2-3, the levels for nonconception and conception cycles are much higher for the Chicago women. The levels for the Bolivian women would have been predictive of inability to conceive in U.S. fertility clinics, but clearly that was not true. In fact, women in this population between ages 20 and 30 had an average of four live births. Given that they were breastfeeding for one to two years, this is not a low birthrate.

One explanation for these differences appears to be diet and physical activity. Consistently, high energy intake seems to be related to high hormone levels, as reported in a number of studies.[12] This phenomenon has also been observed in great apes in whom ovarian hormone levels fluctuate with fruit availability and the energy required to find food. For example, primatologist Cheryl Knott reports that

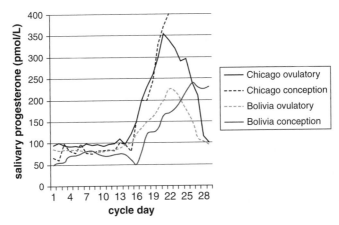

Figure 2-3 Progesterone levels in conception and nonconception cycles in two populations (data from Vitzthum et al., 2004; figure from Trevathan, 2007). Reprinted, with permission, from the *Annual Review of Anthropology,* Volume 36 ©2007 by Annual Reviews, www. annualreviews.org.

among the orangutans she studied in Indonesia, when fruit was abundant and travel to find food was low, ovarian hormone levels were high and conception rates seemed to be higher.[13]

Virginia Vitzthum developed a hypothesis to explain these findings. Her flexible response model (FRM) describes how ovarian function can respond to short-term and long-term fluctuations in resources to delay or speed up reproduction, depending on current ecological circumstances.[14] She argues that if resource limitations are chronic, progesterone levels will be low, but a woman's reproductive functioning will adapt to the low levels, even if it may take a bit longer for her to conceive.[15] On the other hand, if energy is abundant, progesterone levels will be set at a higher level. Short-term energy stress in a woman from a health-rich population can result in a decrease in the levels of progesterone, which may inhibit ovulation, even if her levels are higher than seen in women from health-poor populations. Acute stress can be due to nutritional deprivation (for instance, through dieting) or excess energy expenditure (through heavy exercise). If true, Vitzthum's flexible response model may help to explain why women in health-rich economies experience disrupted ovarian cycles with acute, short-term stresses whereas women from health-poor economies who experience long-term nutritional stress and daily hard labor do not show these disruptions.[16]

Modernization has been shown to have effects on energy budgets and associated effects on reproduction. For example, anthropologists Mhairi Gibson and Ruth Mace studied a population in Ethiopia and noted that birth rates increased after a new water supply system was installed. Previously women had walked several miles a day carrying heavy loads of water. Thus, their energy budgets did not have much left over for reproduction. After the new water system was installed, their workloads declined significantly and they experienced shorter birth intervals.[17] Other researchers[18] found similar fertility changes in a Mayan population they studied when electricity and motorized transportation were introduced. With labor-saving devices came lower energy expenditure and higher birth rates. This phenomenon may partially explain rising fertility rates in populations undergoing modernization.

Along with the attention to the effects that diet can have on reproduction, there has been concern about specific dietary components like phytoestrogens that may influence the reproductive system.[19] A recent study by primatologist Melissa Emery Thompson and her colleagues found that a fruit (*Vitex fischeri*) commonly eaten by chimpanzees in Gombe National Park had the effect of significantly increasing progesterone levels over normal (there were no observed effects on estrogen levels).[20] Although similar effects have been observed in humans (extracts from similar fruits have been used to treat various reproductive "disorders" in humans), it is unclear how widespread are the general effects of phytoestrogens on human reproductive physiology, nor is it known whether there are effects on fecundity in chimpanzees.[21] It is probably safe to say, however, that specific dietary components like phytoestrogens, in addition to overall dietary composition, have an effect on reproductive physiology in humans.

To a great extent, reproductive hormone levels are set during development and reflect resource availability while the girl is growing. As we will see with regard to fetal development, a maturing system (the reproductive system in this case) reads cues about the environment to assess future conditions and adjusts levels of hormones and other components of the system to match the expected conditions.[22] If a girl develops in a health-rich environment, her system "expects" that environment to be stable, but if she experiences short-term deprivation of food, for example, her reproductive system may down regulate to wait for the expected better times. When (and if) they come, full ovarian functioning will likely return. Good evidence of this is seen in the women who tried to conceive during the Dutch famine in World War II but who had much higher rates of infertility and spontaneous abortion compared with the same population before and after the famine ended.[23] For girls growing up in health-poor environments, however, nutritional and work stress may be chronic and "as good as it gets," so their systems adapt and they are able to successfully reproduce under conditions that would lower fertility in health-rich populations. As Vitzthum notes, "the reproductive functioning of women born and living in arduous conditions is not analogous to that of athletes or dieters . . . in wealthier populations."[24]

Studies of migrant populations, most especially those who move from health-poor to health-rich conditions, provide evidence of the effects of timing during development on the female reproductive system. For example, anthropologists Alejandra Nuñez-de la Mora and Gillian Bentley report that Bangladeshi migrants who move to the United Kingdom when they are children have higher progesterone and earlier menarche than those who migrate at or after adolescence or those who remain in Bangladesh. In fact, the early migrants have progesterone levels equivalent to those of native British women, whereas those who migrated near adolescence have levels that look like the women who remained in Bangladesh.[25] This is another illustration of the flexibility of the reproductive system in response to improved conditions as predicted by life history theory.[26]

Several implications of the findings of population variation and short-term changes exist for reproductive hormones. One is that changes in ovarian functioning in the context of nutritional deprivation or heavy exercise should be seen as the system behaving adaptively rather than being the result of a disease or disorder.[27] From an evolutionary perspective, the female reproductive system is attuned to environmental conditions and makes necessary adjustments that increase reproductive success in the long run. In contrast to the view that a certain level of steroid hormones is "normal" and other levels are too high or too low, Ellison argues that the way that ovarian function works is as a "graded continuum" which reflects developmental and ecological circumstances. In between "fully competent cycles" and complete amenorrhea are states of follicular or luteal phase "insufficiencies" and highly irregular cycles, states that may require medical interventions in some contexts but may be expected and normal under other conditions.[28]

Knowing something about population variation in hormone levels throughout the menstrual cycle may also enhance understanding of the elevated risk for some cancers that seem to be influenced by high levels of steroid hormones. Anthropologists Grażyna Jasieńska and Inger Thune summarize this evidence in an overview of breast cancer rates and mid-luteal phase progesterone levels for five populations. The lowest rates and levels for both are seen in the Congo, followed by Nepal, Bolivia, and Poland, with the highest breast cancer rates and progesterone levels in the United States[29] (see Figure 2-4). The levels of progesterone were shown to be associated with total energy intake for each population.

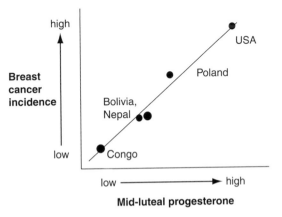

Figure 2-4 Breast cancer rates and mid-luteal progesterone. Adapted from Jasienska and Thune, 2001.

The reported variation in ovarian hormones also encourages closer evaluation of oral contraceptives for women who have different hormonal profiles from those of American and Western European women, on whom most of the research and testing of birth control pills are done. Vitzthum suggests that this can account for the frequently reported difficulties women in health-poor populations have with oral contraceptives.[30] Gillian Bentley, noting that women are adapted for a specific level of steroid hormones to which they were exposed during fetal development, argues that those adapted to low levels have more difficulty metabolizing the hormones in contraceptives designed for women in health-rich populations who experience high endogenous levels.[31] The differences in metabolizing ability may also relate to different lengths of the follicular and luteal phases of the cycle and different duration of menstrual flow. A one-size-fits-all approach to the manufacture of hormonal-based contraceptives puts women in many populations at risk for unpleasant and unacceptable side effects or other negative, and potentially worse, health risks. It is important to note also that hormone levels in the menstrual cycle vary with age,[32] so an oral contraceptive designed for women in their most fertile years may not be appropriate for those at the beginning or end of their reproductive years (Figure 2-5).

With all this talk about higher hormone levels in women from health-rich nations you might wonder if something similar is going on with men. It makes sense that if diet and low activity levels raise reproductive hormones in women, they should do the same in men who have equivalent diets and low activity levels. Indeed, there is evidence that testosterone levels are higher in young men from health-rich populations, but by the time they reach their 60s, the differences are minimal. Similar to what we see in women, the high testosterone levels in young men parallel the high levels of prostate cancer in health-rich nations.[33] There is also

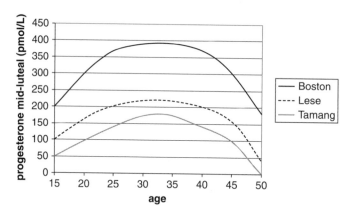

Figure 2-5 Lifetime variation in progesterone levels for three populations. Redrawn from Ellison, 1994, figure from Trevathan, 2007. Reprinted, with permission, from the *Annual Review of Anthropology*, Volume 36 ©2007 by Annual Reviews, www.annualreviews.org.

a suggestion that high rates of prostate cancer in some populations (such as African Americans) may be due to fetal exposure to high levels of maternal hormones.[34] High levels of testosterone can suppress immune function,[35] providing evidence that men, too, experience trade-offs with regard to reproduction and survival. But as with female reproductive hormones, the payoff for reproductive success early in life may be of greater benefit than the risk of suppressed immune function or prostate cancer later in life. To the extent that testosterone is related to mating effort (for instance, in building muscle mass), men in health-rich populations where exposure to infectious disease is low can "afford" to have increased testosterone levels. In recent years there has been a tendency to use "testosterone replacement therapy" for enhancing muscle development or libido, but it is clear that there are risks to the approach just as there are risks to HRT for women.

Should We Reduce the Number of Menstrual Cycles a Woman Has in Her Lifetime?

The mismatch between evolved dietary needs and modern diets seems somewhat easy to resolve, at least in theory if not in practice. Few would argue that there is a downside to reducing fat intake, increasing consumption of complex carbohydrates, and increasing exercise. Trying to bring women's reproductive bodies back in line with ancestral conditions is not as easy to effect, however, because it involves altering reproductive patterns and interfering with hormones. To wit, there are very few popular books written about ways of reducing breast cancer rates by "returning to" reproductive patterns of our ancestors.[36] The point has not been missed by the popular imagination, however, and websites have sprung up advocating ways of altering menstrual patterns to more closely mimic the hormonal milieus of ancestral women.[37]

A lot of women complain about menstruation and some consider it a nuisance and an interference in their social and professional lives, so they have welcomed recent interventions designed to reduce the number of cycles in a year, known as menstrual suppressing oral contraceptives (MSOCs). Arguably, if a woman is already using oral contraceptives that alter normal menstrual function, it does not seem to be too much of a leap to use MSOCs. Published surveys from various parts of the world suggest that many women would choose to reduce the number of menstrual cycles if there were safe ways of doing it.[38] But anthropologist Lynette Leidy Sievert has also found that there are many cultural and personal identity reasons that women choose to maintain "normal" monthly menstrual periods.[39] To understand this, it is important to point out that menstruation has ritual and symbolic meaning in many cultures that have been studied by anthropologists with a common link to fertility and power.[40]

Some women do choose to use MSOCs for protection from pregnancy and to reduce the inconvenience and occasional distress of cycling. Although the health risks do not appear to be much greater than for regular oral contraceptives, Sievert notes a couple of additional concerns.[41] One is that if a woman does get pregnant using MSOCs, she may not realize it for two to three months instead of the two to three weeks that would be signaled by monthly periods. If she engages in behaviors that are risky to a developing fetus (smoking, recreational drug use, alcohol consumption, and poor diet), that two to three months of delayed awareness may result in compromised fetal health and development. Additionally, the quality and quantity of menstrual blood may indicate infections, so if menstruation does not occur very often, infections may be missed. Finally, if women are choosing to use MSOCs because they think they are emulating the ancestral condition, they are quite wrong. MSOCs work with very high levels of estrogen, which is not at all the hormonal milieu of women who are pregnant and breastfeeding for much of their lives.

Most suggestions for using insights from evolution to reduce breast cancer rates have been pharmacologic or surgical[42] rather than lifestyle modifications. This would be analogous to using cholesterol-lowering drugs, gastric bypass surgery, and blood pressure medication to regain health resulting from nonancestral diets, therapies that are not usually considered to be "returning" to an ancestral lifestyle. Some researchers[43] argue that "interventional endocrinology" to reduce breast cancer rates is no more or less "natural" than chemotherapy, radiation, and surgery, especially for those at high risk. They also note that there is little to no resistance today to altering the hormonal milieu with oral contraceptives. But certainly there are more acceptable ways of using what we know about ancestral lifestyles to reduce breast cancer rates through increasing exercise, reducing fat and overall caloric intake, consuming moderate amounts of (or no) alcohol, and increasing dietary fiber.

As evidence of the effects of lifestyle changes on cancer rates, a study of 2,622 former athletes and 2,766 nonathletes showed that athletes had significantly lower lifetime occurrence of breast, ovarian, and endometrial cancers.[44] Most of the athletes had begun training in their teens and were leaner than the others and they probably had lower levels of circulating reproductive hormones. The suggestion that habitual activity may lower estrogen (estradiol) levels was confirmed in a study by Grażyna Jasieńska and her colleagues.[45]

Menstruation Is Rare among Mammals

Most of the review of menstrual cycling so far has focused on the consequences of highly frequent hormonal fluctuations for women's health. A related question is why women menstruate at all, particularly with copious quantities of blood.[46] All female primates of reproductive age have ovarian cycles, but not all produce the blood flow of menstruation as the endometrium is shed following an ovulation that does not result in conception. Rather, as with most mammals, the tissue is reabsorbed. What accounts for this difference? Biologist Margie Profet developed an explanation from an evolutionary perspective in a proposal that menstruation evolved as a way of ridding the uterus of pathogens introduced by sperm during sexual intercourse.[47] The greater amount of menstrual blood shed by humans relative to other primates she attributed to the greater frequency of sexual activity throughout the menstrual cycle of humans. Anthropologist Beverly Strassmann challenged this proposal citing evidence that pathogen load is the same before and after menstruation,[48] and she suggested that the potential for infection is greater at the time of menses because blood is an excellent environment for bacterial proliferation. Profet's hypothesis is also weakened when the relative rarity of menstruation in the past is recalled, especially in the absence of evidence that sexual activity was equivalently rare.

As we have noted frequently, energy conservation is an important component of the evolutionary process. It seems "expensive" to build up the endometrium each month as a preparation for implantation, so it might be expected that "saving" the endometrium from month to month until a conception occurs would be better than tearing it down and rebuilding it every month. Strassman[49] provides evidence, however, that it is metabolically more expensive to maintain the endometrium than to tear it down and rebuild each cycle. Menstruation is associated with iron loss,[50] a risky proposition for women whose iron levels are often compromised by poor diet and intestinal parasitism. But again, as Strassman reminds us, highly frequent menstruation and associated potential iron loss was somewhat rare in the human evolutionary past. This same argument has been proffered to explain why the negative consequences of menstruation manifested in what has come to be called PMS (premenstrual syndrome) were not likely of concern in the past, which will be discussed later.

Not every interesting phenomenon related to human reproduction needs to be explained as a product of natural selection itself. An alternative view from those described earlier is that menstruation in humans simply co-evolved as a by-product of the complex reproductive system humans have and its existence does not really need explaining.[51] A general principle in science, known as "Ockham's Razor" is that the simplest explanation is the best until proved otherwise. Perhaps the by-product argument that menstruation occurred along with the deep invasion of the placenta in humans is an example of this principle.[52]

The Story of the Egg, Part I

The story of the human egg (ovum) begins at about 20 weeks of gestation, when 4–6 million "pre-eggs" (oogonia) develop by the process of mitosis, yielding cells with 46 chromosomes each. By about the sixth month of pregnancy, these oogonia begin the first stage of meiosis, which reduces the chromosome number to half, in preparation for combining with sperm many years later to produce an embryo with 46 chromosomes. The result of this first stage of meiosis is called a primary oocyte and it will not undergo any further cell division until puberty, a phenomenon that has been called a "kind of suspended animation."[53] The oocytes will remain in this state for up to 50 years, housed in the ovary, and there will never be any more of them than there are at this point, partway through fetal development. In fact, their numbers begin a steady downhill fall[54] until there are just a few left at the time of menopause. The rate of loss may be as high as one oocyte every four minutes.[55] This may seem wasteful, but given that the vast majority of the oocytes that are lost are defective, it is actually an example of the reproductive tract acting as an agent of natural selection to eliminate conceptions that are not likely to result in live births or healthy offspring. As we will see, one proposal for why ovulation ceases altogether at menopause is that the ovary has simply run out of eggs (see Chapter 9). Given all the insults that the eggs can experience in the next five decades, it is perhaps not surprising that chromosomal abnormalities are more common in babies born to older women.

In the embryonic ovary the oocyte is housed in a follicle that, in parallel with the oocyte, is called a primary follicle and, like the primary oocyte, it remains pretty much unchanged until puberty. At puberty, the oocyte and the follicle undergo change from primary to tertiary. At ovulation, it is the tertiary oocyte that undergoes meiosis (in response to the LH/FSH surge at midcycle) and is ovulated from the tertiary follicle. The oocyte is released into the abdominal cavity from which it is sucked into the fallopian tube. For the next 10–12 hours it is available for fertilization by sperm. Meanwhile, back in the ovary, the tertiary follicle that remains following release of the egg becomes the *corpus luteum*. The corpus luteum produces estrogen and progesterone for several days and if the egg is not fertilized, the corpus luteum dies. If the egg is fertilized, the corpus luteum continues to produce progesterone ("pro-gestation") and some estrogen until the placenta takes over this task.

The tertiary oocyte undergoes the next phase of meiosis when the sperm contacts it, at which point it is appropriate to call it an *ovum*. Once it is fertilized, it is called a *zygote* and it travels through the fallopian tubes to the uterus, a trip that lasts about a week. Meanwhile, the corpus luteum is producing progesterone to prepare the uterine wall for implantation. When it implants, the zygote becomes a *blastocyst*, which imbeds in the uterine lining. One part of the blastocyst, the *inner cell mass*, becomes the embryo; the other part, the *trophoblast*, becomes the placenta. At this point, human chorionic gonadotropin (HCG) begins to be secreted by the trophoblast, which can be detected in an over-the-counter pregnancy test. At any point in the entire process, something could go wrong and lead to a failure to ovulate, a failure in the corpus luteum, or a failure to implant. These include insufficient progesterone production, short luteal phase, infectious agents, genetic abnormalities, maternal immune system responses, intrauterine devices (IUDs), and intervention by such agents as RU-486, which blocks progesterone receptors. Chapters 3 and 4 continue the story of the egg after it is fertilized and proceeds through the next nine months of pregnancy to become a human.

During each menstrual cycle, progesterone and estrogen stimulate cell division in the breasts in the second half of the cycle as the breasts and uterus prepare for implantation. When fertilization does not occur, the newly created cells die until the next cycle when the process starts over again. If a woman never gets pregnant, this cell creation–cell dying cycle may occur over and over for more than 30 years. Unfortunately, sometimes the cell division gets out of hand resulting in breast or endometrial cancer, which is why rates of these cancers are higher in women who have never been pregnant.[56]

The "Normal" Menstrual Cycle

Most of us are familiar with what the "normal" menstrual cycle looks like from grade school health classes and high school biology courses. As usually described, the cycle is divided into two parts, a *follicular phase* when the ovarian follicle is developing under the influence of estrogen, and a *luteal phase* following ovulation when the corpus luteum secretes high amounts of progesterone in preparation for implantation of a fertilized egg. The follicular phase is the time from the first day of menstruation until ovulation, typically lasting about two weeks. The luteal phase is the second half of the cycle, from ovulation to menstruation, another two weeks. A "normal" menstrual cycle is about 28–30 days, although there is much variation from woman to woman and for the same woman from month to month.

The fact that I keep putting the word "normal" in quotation marks should alert the reader that what is normal for one population or one woman is not necessarily normal for another. Furthermore, as I emphasized in the introductory chapter, most especially what is defined as normal in the medical literature may be true for well-nourished healthy women in health-rich populations, but the actual range of variation is far greater and a lot of that range is still healthy. This is no less true for our understanding of the "normal" menstrual cycle, in terms of regularity, length, and volume of blood produced. For example, Virginia Vitzthum and her colleagues found that only about a third of the women in their Bolivian sample had cycles that were "regular," the textbook definition of which is having all of their cycles fall in the range of 26–32 days in length.[57] As the authors note, this has important implications for contraceptive use when their success depends on regularity of cycle lengths. It also calls into question the potential success of the rhythm method of birth control in a population where the majority of women do not have 28–29-day cycles.

In this same population, Vitzthum and her colleagues found that the mean duration of menses flow was 3.7 days (median 3.5).[58] This is at the low end of a range of menses lengths reported for numerous other populations. In general, women from health-poor populations have shorter menses than women from health-rich populations. European and U.S. median lengths approach six days, the maximum among populations reported. Vitzthum and her colleagues suggest that, as with many aspects of women's biology I have been discussing, the length may be related to levels of ovarian hormones produced in the cycles: where progesterone and estrogen levels are high, days of menses are longer and where they are low, flows are shorter. Other studies have reported that the relative risk for reproductive cancers is higher for women with long menses (greater than six days compared with fewer than four days in length). These findings led Vitzthum to suggest that another reason some contraceptive options may not be acceptable to women with short menses is that they will not want to lengthen them into what they may regard as "abnormal" cycles.

Problems with Menstrual Cycling: Inconvenience, Pain, and PMS

Almost every woman has experienced cramping during menstrual periods and varying degrees of cycle irregularity. None of these is particularly enjoyable, and for some women, cramps can be debilitating. How did women in the ancestral past keep up with the demands placed on them for gathering food and caring for children when they had severe cramps, several days of bleeding, or PMS? As a quick reminder, even if they routinely had problems during menstruation, those periods were probably few and far between, so they may not have taken an excessive toll on their normal daily activities.

Premenstrual syndrome is seen as a routine and expected part of being of reproductive age for women in the United States and many other populations, although its existence as a universal phenomenon for all women is questioned because it is rarely mentioned in the traditional anthropological literature. Of course that could be because it was not considered a subject worthy or appropriate of research and anthropological inquiry in the past, but it nevertheless appears to be more common in health-rich populations and is sometimes considered to be a "disease of civilization."

What is PMS? One of the problems with characterizing PMS is that there are so many "symptoms" listed as being associated with it that defining it is almost impossible. It is often defined as a cluster of symptoms that occur a few days before menstruation. Here is a partial list of experiences that a woman might label as PMS: weight gain, breast pain, swelling, dizziness, cold sweats, nausea, headache, backache, hot flashes, general body aches, sleeplessness, forgetfulness, difficulty concentrating, feeling uncoordinated, poor performance at work or school, food cravings, mood swings, crying, depression, anxiety, irritability, hostility, tension, loss of interest in sex.[59] So if PMS is anything that happens in the few days before menstruation then it can become a self-fulfilling prophesy in that any negative physical or emotional sensation can be labeled PMS. It is estimated that as many as 70% of women in the United States suffer from PMS by this broad "cluster of symptoms" definition. For others, however, the experience is so stressful and debilitating that it may be more appropriately labeled premenstrual dysphoric disorder (PMDD); fewer than 8% of American women have symptoms this severe.[60]

The history of research and thinking about PMS has generally not been very complimentary of women.[61] For example, it has variously been ascribed to women who were not satisfied with their roles in life, were neurotic and tended to sicken from emotional stress, had too much time on their hands, and to those who had trouble becoming pregnant or who were dissatisfied with their relationships. The uterus was believed to be the root of all problems that women had, so it was blamed for PMS as well.[62] More recently, there has been a push toward accepting it as a medical diagnosis and researching medical ways of alleviating it. PMDD and a number of related disorders are now classified in the *Diagnostic and Statistical Manual of Mental Disorders* of the American Psychological Association, giving it a more accepted status.

Unfortunately, by calling PMS a medical disorder, a common aspect of reproduction, which usually proceeds with little notice and certainly no need for medical intervention, has become pathologized. The debate about whether the causes of PMS are hormonal or psychological continues. Current thinking concludes that PMS has a biological basis but is exacerbated by current and past lifestyle stresses. Additionally, it has been argued that women who experience it may have higher sensitivity to normal hormonal shifts.[63]

As noted, the fact that PMS seems to be much more common among affluent women could be due to the relatively high levels of ovarian hormones that they have in comparison with women in health-poor populations. Perhaps the drop from high progesterone levels at the end of a cycle is much more precipitous for affluent women and thus the "withdrawal" of hormones is much more acutely felt, triggering the various physical and psychological sensations that come at the end of the cycle.

As another way of thinking about the "highs" of the ovulatory period, anthropologist Chris Reiber hypothesizes that the fall from high levels of flirtation, good mood, self-confidence, and sociability that often accompany the ovulatory period is experienced by women as low mood, low self-confidence, and low levels of sociability and perceived as PMS.[64] He argues that PMS is the inverse of the heightened social and behavioral aspects of mate attraction during the most likely time to ovulate. In fact, the most noticeable aspect of the premenstrual period for women is that they simply do not feel as good as they did a week or two earlier.

Reiber continues his assessment of PMS from an evolutionary perspective by suggesting that under conditions not favorable for reproduction, women do not feel the heightened emotions and behaviors at the time of ovulation and they do not "fall" from those heights pre-menstrually. Thus, they are not likely to experience the sensations associated with PMS. By this reasoning, he argues, PMS occurs when women are living under conditions favorable for reproduction, which would explain why it is more common in affluent populations. He finds support for his proposal in data showing that women for whom conditions for reproduction are not favorable show negative moods and behaviors at the time of ovulation and more positive mood pre-menstrually. This also may explain why women experience varying levels of PMS (or not at all) at various times in their lives. This perspective helps us understand the amorphous and variable symptoms of PMS and why medicine has been so unsuccessful in treating those symptoms. Furthermore, the highs at ovulation may be higher in populations that have high levels of ovarian hormones, resulting in a more precipitous fall, as I have noted.

Suppressed Immune Function during the Luteal Phase

One of the most promising areas of evolutionary thinking about PMS relates to its occurrence in the phase of the menstrual cycle when immune function is most compromised. As we will see in the next chapter, this suppression in the period following ovulation enhances the likelihood that the mother's immune system will not reject a fertilized zygote and it will successfully implant in the uterine wall. Although anecdotal reports of greater susceptibility to illness during the luteal phase of the menstrual cycle are numerous, there have been very few well-controlled studies and several have produced conflicting results. Evidence that the effectiveness of mammograms is enhanced if done during the follicular phase[65] and that surgery for breast cancer may contribute to better survival when done during the follicular phase[66] lends support to the idea that hormonal variation during the menstrual cycle affects aspects of health other than reproduction. A number of diseases and disorders are known to be associated with increased symptom reporting pre-menstrually (during the late luteal phase) and include asthma, migraine headaches, diabetes, arthritis, depression, gastrointestinal diseases, osteoporosis, chronic fatigue syndrome, and epilepsy.[67] On the other hand, some autoimmune disorders like multiple sclerosis and rheumatoid arthritis seem to be ameliorated during the luteal phase, but lupus worsens.[68]

An early study of 10 healthy Mexican women showed that an immune factor was significantly higher during the follicular phase for each woman than during the luteal phase, suggesting that this first line of defense against potential pathogens is compromised during the luteal phase.[69] Studies of immunosuppression in pregnancy suggest a role for progesterone as an inhibitor of immune function. For example, one immune factor showed decreased levels during the luteal phase of the menstrual cycle in the presence of progesterone, suggesting initial suppression of the immune system in preparation for implantation of an embryo.[70] Levels of another immune factor are inversely related to progesterone levels.[71] Monkeys treated with estrogen show increased resistance to intravaginally administered SIV (a monkey virus similar to HIV), whereas those treated with progesterone show extreme susceptibility.[72] Progesterone is reported to increase susceptibility to *Chlamydia* infection in rats, whereas estradiol decreases susceptibility.[73] On the other hand, estrogen in the luteal phase was shown to lower resistance to *Candida albicans* in an experimental study, whereas progesterone had no effect.[74]

In another study,[75] women recorded fewer symptoms in the follicular phase and fewer illness onsets during the menstrual (early follicular) phase, suggesting that immune response to viral and bacterial infections may vary throughout the menstrual cycle. In our own analysis of 183 menstrual cycles, significantly more illnesses began during the luteal phase than during the follicular phase, although the total number of days of illness was higher in the first half of the cycle.

Biologist Caroline Doyle and her colleagues hypothesize that some symptoms attributable to PMS are the result of infectious agents that have greater impact during the luteal phase when immune response is dampened.[76] Furthermore, they suggest that appropriate treatments for these aspects of PMS would be anti-infective drugs that target the causative agents rather than analgesics and similar efforts to relieve the symptoms. Paul Ewald, one of the coauthors of the Doyle paper, is a proponent of seeking infectious causes of chronic diseases and he suggests that focusing on premenstrual exacerbation of certain chronic diseases may help elucidate their causes.

Ewald extends this argument to critique the hypothesis that high levels of ovarian hormones are related to reproductive cancers. He suggests that the "hormonal proliferation hypothesis" for the increased rates of breast cancer is insufficient to explain the great majority of cases. He focuses on the immune suppression effects of estrogen and progesterone rather than on their direct effects on cell proliferation and proposes that when the levels of these hormones are high, immune responses are suppressed so that a woman becomes vulnerable to a number of infectious agents.[77] He cites viruses (including Epstein Barr/EBV and Human Papilloma Virus/HPV) that are linked to breast cancer. Noting that risk of breast cancer is high during pregnancy, he suggests that this may be due to the immunosuppressive effects of progesterone. For women in health-rich populations with elevated lifelong levels of estrogen and progesterone, the higher risk of breast cancer may result not only from independent lifestyle factors and underlying genes but also from the potentially greater effects of the hormones on immune function. I would not be surprised if it is one day determined that all of these factors, infectious and noninfectious, are contributing to increased cancer risks.

Menstrual Synchrony

One of the most compelling ideas to come out of research on the menstrual cycle in recent years is that women who live or spend a lot of time together synchronize their cycles after a few months. Unfortunately, it is probably a myth. When menstrual synchrony was first reported by Martha McClintock in an article in the esteemed scientific journal *Nature*,[78] it was enthusiastically embraced by women who not only found the concept attractive but had actually experienced it themselves. What McClintock described and many women reported anecdotally is that after a few months of living together young women find that their menses onsets gradually converge so that they are cycling together.[79] Dozens of attempts to replicate McClintock's findings followed, with about half finding support for synchrony in their samples[80] and about half not finding support.[81] These were followed by methodological criticisms that called into question the findings of purported synchrony.[82]

The phenomenon of synchronized ovulations has been well established in rats and many other species. Unlike most mammals, however, humans do not seem to be influenced by olfactory (chemical) signals known as pheromones, even in courtship and reproduction, times when they play major roles for many species. Although some scientists are unconvinced of their existence in humans,[83] there have been suspicions that we are not entirely free from pheromones, and menstrual synchrony was believed to provide evidence of that. Several studies tested the hypothesis that synchrony was driven by pheromones using underarm secretions believed to contain the chemical compounds that drove the cycles.[84] These studies were also severely criticized on methodological grounds.[85]

Those who were convinced that the potential for menstrual synchrony existed agued that various environmental factors interfered with its expression such as exposure to men, work-related stresses, the degree of intimacy between the women, and breastfeeding. Based on the idea that close physical proximity between women and little exposure to men would be the best circumstance under which synchrony would emerge if it exists, psychologist colleagues and I searched for evidence of synchrony in a sample of 29 lesbian partners who slept in the same bed, were sexually intimate with each other, and had no intimate contact with men. We found no evidence of menstrual synchrony in three cycles of these women and, in fact, more of their cycles diverged than converged over the three cycles.[86] In order to converge, cycles of two women that are initially variable and different would have to adjust so that they are the same length and show decreased or at least equivalent variability. Menstrual cycles of contemporary women who are not on oral contraceptives are nothing if they are not variable, so this circumstance would be highly unlikely, no matter what the pheromonal conditions.

Menstrual synchrony has always been difficult to explain from an evolutionary perspective because of the now well-known fact that most women in the past experienced very few menstrual cycles in their lifetimes and rarely cycled long enough for synchrony to be established. Furthermore, group size was small (25–40), limiting the number of women who were neither pregnant nor lactating at any one time. Perhaps the only time in life when women had a series of consecutive menstrual cycles was in the first two to three years following menarche and the last two to three years before menopause when most cycles are highly irregular and often anovulatory. In fact, any women who cycled frequently enough for synchrony to be established were probably sterile or subfecund, so they were unlikely to pass along characteristics that affect the ability to synchronize.

Furthermore, if synchrony has an evolutionary basis via an effect on reproduction, it is ovulation synchrony that is the relevant event. In addition to collecting data on menses onset for the 29 lesbian couples in our study, we also collected data on basal body temperature that could be used to infer ovulation dates. We found no evidence of ovulation synchrony. We and others have concluded that there is no evidence that menstrual or ovulatory synchrony are normal occurrences in contemporary or past populations, but this does not mean that there are no general effects of sociosexual behaviors on reproduction in humans.[87]

Human Female Sexual Behavior: Is There a Hormonal Influence?

The behaviors of most mammals are influenced by the levels of reproductive hormones they experience during the reproductive years. Most notably, sexual behavior tends to be concentrated in the period when ovulation occurs and neither females nor males are as interested in sexual interaction at other times during an ovulatory cycle or during pregnancy. Many primate species have sex at times other than when fertilization can occur, but humans seem to take this to an extreme. Anthropologist Helen Fisher [88] has called us "sex athletes" because of our apparent interest in engaging in sexual activity frequently and at times other than when we can reproduce.[89]

In most nonhuman mammals, females make it clear to other members of their species that it is time to copulate. Species do this in at least three different ways (with many, sometimes extravagant, variations): they exhibit chemical signals, or pheromones that indicate their reproductive status by odor; they have external visual signs like the red, swollen buttocks region of baboons and chimpanzees; or they engage in various behaviors that show their willingness to copulate. These things happen in most primates at the time when a female is most likely to conceive (the time of ovulation) so that for most species, sex is linked very closely to reproduction. Presumably for most species, there are disadvantages to having sex when there is low likelihood of pregnancy. These disadvantages could include energy expenditure, competition among males, vulnerability to predators, and risks to females with young infants. Natural selection has acted in those species (indeed in most species) to confine sexual behavior to the time when likelihood of reproductive success is greatest.

For women, sex and reproduction are not necessarily linked. Women today can and do have sex at any time during their monthly cycles, whether they are likely to get pregnant or not. Some cultures impose rules about when during a cycle women can (or, more often, cannot) engage in sex, but there are no obvious biological limitations on the timing of sexual behavior. In fact, not only do men not know when ovulation occurs but women themselves usually do not know.[90] There are techniques that can be used to determine time of ovulation (either in order to maximize or to minimize chances of getting pregnant) such as measuring basal body temperature or examining cervical mucus, but the proliferation of home ovulation predictor kits indicates just how difficult it is for most women to pinpoint when this important part of their reproductive cycle occurs. At some point in our evolutionary past, human females "lost" estrus and began to conceal ovulation. It makes sense to think that these events were related and that they led to the increase in what we can call nonovulatory sex or nonreproductive copulation that we see in humans today.

Not only are women sexually active throughout the ovulatory cycle but they also continue to have interest in sex at times when conception is unlikely or even impossible, such as after menopause, during pregnancy, and during lactation. Again, some cultures have rules about when sexual activity may resume following pregnancy but the existence of these rules confirms that from a biological point of view, there is nothing to prevent women from participating in sexual activity during this time of very low fecundity. In fact, there is evidence that in the first few years after menopause, many women are *more* interested in sex because of freedom from worries about unwanted pregnancy. This suggests that in humans, sex serves a function beyond the narrowly reproductive one and that for women, there must be some long-term advantage to having sex even when pregnancy is not likely. In evolution, advantages are always ultimately reproductive, so this means that in the long run, women who had sex even when they weren't capable of becoming pregnant were more reproductively successful than women who had sex only during ovulation.

Despite the evidence that women cannot easily judge when they are ovulating, there is some evidence for minor changes in women's behavior and attractiveness to others at the time of ovulation, suggesting that hormones continue to have an effect on our interest in sex and the interest of those around us in having sex with us. Although results are often conflicting, many researchers have reported significant phase-related peaks in sexual activity, desire, and arousability in human females.[91] One problem with conflicting studies is that it "takes two to tango," and patterns of sexual activity may be as much a reflection of a partner's interest as a reflection of a woman's own libido. Certainly partner availability has a lot to do with sexual behavior as well.

The prediction is that if there were remnants of ancestral sexual activity patterns, we would see slight elevations at the time of ovulation. Indeed, a number of studies have demonstrated just that, finding increases of sexual interest and activity in the few days before ovulation, during the late follicular phase.[92] The high levels of progesterone during the luteal phase have often been implicated in the lowered sexual interest following ovulation and preceding menstruation. Indeed, in research that I conducted with colleagues in psychology, we found that sexual behavior with or without a partner reached its lowest point in the early luteal phase for most of the groups studied.[93] Thus, it may be that progesterone has a dampening effect on women's libido, just as it does on immune function. In fact, the two may be related. If a woman is less likely to engage in sex during the luteal phase, she reduces her exposure to potential infectious agents that can come from sexual intercourse.

Given that this is the time for implantation of a vulnerable embryo, lowered libido may serve a protective function. In the anovulatory cycles of the women we studied, there was no evidence of decreased sexual activity during the last half of the cycle, providing support for this proposal.

Entirely within the realm of speculation, I tentatively suggest that the familiar notion of higher levels of ovarian hormones affecting behavior and health may also manifest by influencing levels and patterns of sexual activity. An intriguing study of wild chimpanzees reports that in cycles with higher estrogen levels there were more copulations and more conceptions than in cycles with lower levels, most likely due to greater attractiveness to males.[94] Although the factors affecting human female sexuality are too numerous and complex to discuss here, I am willing to speculate that the high levels of ovarian hormones found in women in health-rich populations have an effect on anecdotally reported higher levels of sexual activity in these populations. This may also explain the lowered interest in sexual activity reported by pregnant women and those nearing menopause. And as with other aspects of behavior affected by the fall from high levels at ovulation and the higher progesterone levels in the luteal phase, it may be that women from health-rich populations are more likely to show evidence of cyclic variation in sexual behavior.

A far-reaching conclusion about population variation in ovarian hormone levels is that the high levels reported for women in health-rich populations are actually at one end of a continuum, an end that lies outside the range that probably represents most of human evolutionary history. As anthropologist Susan Lipson says, the levels that we see today and regard as "normal" may result from what happens when "the energetic constraints under which the system was designed to operate have largely been removed."[95] Unfortunately, what we get for that is early menarche and increased cancer risks, among other things. In fact, as I argue throughout this book, this one factor may account for many of the health "problems" women face during their lives. Understanding this and doing something about it, however, are very different issues.

This chapter has been about menstrual cycling focusing on those cycles that do not result in pregnancy. Fortunately, for reproductive success, some menstrual cycles result in conception and proceed to pregnancy, although it is estimated that as many as one in four conceptions do not make it.[96] Slightly more than half of these are lost in the first 9–10 days, often before a woman is aware she has conceived. Most of the rest are lost within the next month. As we will see in the next chapter, pregnancy loss is of great medical and personal concern, but it may be adaptive and thus naturally selected under some circumstances.

3

Getting Pregnant: Why Can't Everyone Just Get Along?

How easy is it to get pregnant? Apparently, it is too easy for some but very difficult for others. In studies conducted in health-rich countries, the average amount of time it takes to get pregnant (called "waiting time to conception") is seven to eight months. At maximum fertility, in her early 20s, a woman has only about a 25% chance of getting pregnant per unprotected cycle, and it has been estimated that it takes about 100 acts of intercourse per pregnancy.[1] Why should it be so difficult? As we will see later, highly frequent sexual activity for a couple may play an important role in making sure her immune system does not attack the fertilized egg. But beyond that, other factors that seem to affect fertility include age, health (such as presence of sexually transmitted diseases [STDs], or diseases like malaria), genetic incompatibilities, exposure to ubiquitous biochemicals in our environment, and lifestyle factors like smoking and alcohol and drug use. In this chapter I will focus primarily on aspects of evolved biology that impact pregnancy, although the other proximate factors, including sociocultural variables, play equally important roles.

Polycystic Ovaries as a Cause of Infertility

One frequent cause of female infertility is polycystic ovarian disease or syndrome (PCOD or PCOS), which seems to have a genetic basis and is associated with obesity and a variety of hormone imbalances, including those that underlie conditions that predispose a person to heart disease and diabetes (known collectively as "metabolic syndrome X"). Given that these seem to be on the increase worldwide, it is likely that PCOD will also increase in the future. Women with PCOD produce a large number of very small follicles that typically fail to ovulate.[2] In many women it is not a permanent condition and in cases where it is accompanied by obesity, a small weight loss will often result in ovulation. Some scholars have suggested that it can be regarded as a "fertility storage condition" that allows fertility to be delayed until physical and environmental circumstances are more supportive of reproduction.[3]

Other researchers have noted that increased rates of PCOS parallel rates of obesity and insulin resistance and are particularly common in populations that are undergoing lifestyle transition or "Westernization," and suggest that the syndrome is a result of entering into an evolutionarily novel environment.[4] According to this argument, this sort of transition puts women at risk for a number of challenges to reproduction and overall health, suggesting that we may begin to see increases in rates of reproductive failure as more and more people migrate to areas of relatively greater resource abundance and lower levels of physical labor.

Another view proposes that the elaboration of the endometrium necessary to support a large-brained fetus has left humans vulnerable to disorders such as PCOS.[5] In particular, humans have unusually long follicular phases, presumably required to provide time and sufficient hormone production to form the more complex endometrium into which the placenta will implant if conception occurs. But this prolonged follicular phase makes humans vulnerable to excessive hormone production and to ovulation failure. Cell turnover rates are also higher in women with PCOS, which may explain their greater vulnerability to endometrial cancers. This proposal is in agreement with the argument that PCOS may have been adaptive at one time when caloric intake was lower but that today, in richer environments, it is a liability.

Evidence that women with PCOD have delayed menopause suggests that not only do their more numerous follicles last longer, but they may be able to make up for lost fertility early in life by reproducing later than usual if conditions improve.[6] This concept has been linked with the fetal origins hypothesis, which argues that conditions during fetal development have long-term impacts on later life fertility and health (discussed later in this and the next chapter). Women with metabolic syndrome X and other hormone imbalances during pregnancy may have fetuses that are "programmed" toward similar problems later in life, including insulin resistance, obesity, and PCOS. Whether these actually develop depends on the conditions under which growth and subsequent reproduction occur. Thus, PCOS can be seen as an example of a phenomenon that is problematic under certain circumstances (a "defect") and adaptive (a "defense") under others. This warrants referring to the condition as simply PCO (which may be good) rather than PCOS or PCOD, both of which explicitly emphasize the defect concept.

Physiological Changes in Pregnancy

Many aspects of a woman's physiology are altered when she becomes pregnant. One of the most important is that she metabolizes food more efficiently and gets more nutrients from foods she consumes than she does when she is not pregnant. This leads to weight gain even when food intake does not increase or when it decreases, as it often does early in pregnancy in association with morning sickness. The proximate mechanisms for changed metabolism are hormonal alterations that optimize digestion. Among the hormones that increase is one (cholecystokinin) that also makes a woman feel tired and listless after a meal—such decreased activity also contributes to weight gain,[7] in addition to enhancing energy conservation. Now that she is "eating for two," all of these changes can be seen as contributing to improve gestational environment. This increased ability to put on weight and store energy as fat was clearly adaptive for our ancestors, but in today's world, where food availability is sometimes excessive, this ability raises concerns for women and their health care providers.

Fetal and maternal developmental changes during pregnancy are so complex that it is amazing that any of us get born. If you read enough about what can go wrong, it seems a miracle that any pregnancy makes it to the end without a damaged fetus or a maternal system that can never conceive again. In fact, so much of the medical literature emphasizes the things that can go wrong that women often experience as much anxiety as joy over their pregnancies. An evolutionary perspective recognizes that when things do go wrong, there are often good evolutionary reasons behind it.

In the previous chapter we discussed the development of the egg, noting that its beginnings are during early embryonic development. When reproductive maturity is reached and conception occurs, the mother's system must be able to recognize that fact and work to maintain the pregnancy for the next nine months. Recall that when ovulation occurs, the corpus luteum produces estrogen and progesterone for several days so that if the egg is fertilized, it can be "rescued" and can continue to produce hormones required in early pregnancy, especially progesterone, until production is taken over by the placenta several weeks later. The term *rescue* is frequently used in biology and medical texts because the assumption is that the "normal" (most common) occurrence is for the corpus luteum to deteriorate in women who spend more time having menstrual cycles than not. It assumes that termination of the corpus luteum is the normal and expected occurrence at the end of the menstrual cycle, whereas in the ancestral past, its persistence (due to pregnancy) may have been more common than its demise, given that highly frequent menstrual cycling was relatively rare.

From the time that the corpus luteum is rescued until the baby is born, so many interrelated systems must develop and function right that to describe them is to emphasize what can go wrong. Here are just a few of the things that happen from fertilization onward: rescue of the corpus luteum, implantation of the zygote, burrowing of the trophoblast into the uterine lining, development and proper functioning of the placenta, establishment of communication between mother and fetus through the circulatory systems, adequate nutrition for the mother so that nutrients can pass to the fetus, adequate oxygen transport, hormonal regulation of fetal growth, and keeping the uterus from having contractions until labor begins at term.

To move forward in our discussion of pregnancy, we need to describe some of the hormones that orchestrate the process. The placenta produces a hormone similar to growth hormone, the production of which increases throughout pregnancy. It functions to make glucose more readily available to the fetus and, later, enhances milk production. Levels are proportionate to placental size and both have an influence on how large the fetus becomes. Estrogen, also produced by the placenta during pregnancy, stimulates uterine growth and development of the mammary glands in preparation for later breastfeeding. Progesterone from the placenta is secreted to both mother and fetus and helps to maintain the uterus and breasts and inhibits ovulation. It also affects thirst, appetite, and fat deposition. Clinical guidelines suggest that "normal" progesterone levels in pregnancy are those reported for well-nourished women in Chicago or Boston (see Figure 2-3 in Chapter 2) and that when the levels are too low, complications may result. But anthropologists who study populations that are not as well nourished (overly nourished?) report much lower levels, levels that would, in fact, be deemed problematic in Boston. In fact, pregnancy loss is no greater in women in populations with lower progesterone than in those with high levels, again calling into question the medical understanding of normal.[8]

Pregnancy is usually described in thirds (trimesters) because very different actions are taking place in each and the timing of growth and development events is somewhat constrained. The first trimester is one of growth and differentiation of the embryo, the second is when total length of the fetus increases, and the third is when the developing fetus is putting on weight as fat. Our discussion will examine what goes on in each of these trimesters and how our evolutionary history reflects the changes and impacts the successes or failures.

First Trimester: The Story of the Egg, Part II

In Chapter 2 we followed the egg from its embryonic stage until fertilization and implantation, or, more commonly today, menstruation. Here we will back up a bit and focus on what happens when the zygote prepares to implant into the uterine lining as a blastocyst. As noted in Chapter 2, part of the blastocyst (the inner cell mass) becomes the embryo and part (the trophoblast) becomes the placenta. From the view of clinical medicine, a failure to implant is a problem and a common cause of what seems to be infertility in women. Clearly some cases of failure to implant are pathological and can benefit from medical or pharmacologic assistance, but there are numerous situations in which failure of a pregnancy at this stage is a "good thing" from the perspective of evolutionary medicine.

The zygote that results from the union of egg and sperm is unique and genetically different from both the mother and the father. Our immune systems are designed to deal harshly with organisms that are not familiar by damaging or outright rejecting them. This is fine when the invading organism is a pathogen, but when it is the beginnings of a hoped-for pregnancy, expulsion may not be desirable. The sperm that contributes to formation of the zygote has already successfully survived passage through the cervix, the vaginal mucus, and the fallopian tubes; at any point in its journey to the egg the mother's system could have destroyed it. For a while the zygote can "hide" from the mother's immune system, but eventually it is fully exposed and vulnerable. As noted in Chapter 2, the luteal phase following ovulation is a time when the mother's immune system is slightly dampened. This is a time when she might be more vulnerable to illnesses, but this dampening serves the zygote well because it is less likely to be detected and rejected. This is one of those trade-offs: a woman is a bit less protected from illness during the luteal phase and early in pregnancy, but if she does conceive, she is a bit less likely to reject the zygote.[9]

Implantation is the process whereby the embryo imbeds itself in the endometrial wall of the uterus. As noted in Chapter 2, this is when HCG begins to be secreted by the trophoblast (a sort of pre-placenta) and a simple urine test can reveal the pregnancy. It is probably appropriate to say that this is when pregnancy "officially" begins and it is an important step in that a high percentage of fertilized eggs fail to implant.[7] The ability to secrete HCG serves as a signal to the mother's system that the embryo is viable, at least to this point.[10] The egg and the endometrium work together to coordinate implantation; if the timing is off on either side, the process will likely fail. But if all goes well, the embryo becomes successfully attached to the uterine wall and exchange of nutrients and oxygen begins. If things do not go well, or if implantation is delayed too long, the embryo is likely lost in what the mother perceives as a late, and perhaps especially heavy, menstrual flow. She may not even realize that she had conceived.

It may seem surprising that so many fertilizations and conceptions are lost early in pregnancy. What a waste, and how sad the loss is for so many couples trying to get pregnant. But from an evolutionary perspective, it makes sense that embryos that may not have a good chance of surviving pregnancy to grow up and reproduce themselves are discarded before the mother invests too much time and energy in gestation. As Virginia Vitzthum says, rather than humans having a defective or inefficient reproductive system, it appears that we have one that "has been designed to be flexible, ruthlessly efficient, and, most importantly, strategic."[11] This will lead us to an evaluation of early pregnancy loss that differs from the clinical view,[12] as we will see later.

Once pregnancy is well established, by the end of the second month, the percentage of embryos that are lost drops precipitously. A lot of that is up to the communication system between the mother and infant via the placenta, so before moving on in this pregnancy, let's learn more about this amazing organ.

The Placenta

The largest group of mammals is sometimes referred to as the "placental mammals," illustrating the significance of the placenta in understanding characteristics of this subclass of mammals. There are several types of placentas, reflecting different demands of pregnancy among mammals. One way in which they are classified reflects the number of membrane layers between the maternal and fetal systems and how much of a barrier exists between the two circulatory systems. Humans have a type of placenta that enables fetal tissues to invade deeply into maternal tissues, so that there is only a thin layer between fetal and maternal blood vessels. Furthermore, the deeper invasion provides for a greater surface area over which nutrients and gases can be exchanged. This thin barrier between the maternal and fetal system also allows freer exchange, via diffusion, of molecules, both good (nutrients and oxygen) and bad (drugs, toxins, et cetera).[13] Other mammals with this type of placenta include monkeys, apes, tarsiers, rodents, and rabbits.[14]

Even within a group of mammals with the same type of placenta, there are a number of differences that relate to reproductive strategies of different species. For example, the mouse has the same general type of placenta that humans have but it gestates a dozen or more fetuses for only about three weeks, whereas the human usually carries only a single fetus for nearly 13 times that long. How can the same type of placenta handle such radically different gestational physiologies? One argument proposes that placental genes are differentially expressed in the first and second halves of pregnancy.[15] In the first half, the genes that are charged with getting the pregnancy started and with basic energy and gas exchange are similar in both mice and humans because the demands are similar at this stage. Once that has happened, more recently evolved (and mouse- or human-specific) genes kick in and allow for the short gestation in mice and the long gestation in humans and all of the associated differing physiological requirements and challenges.

The fetus is sometimes referred to as a graft. Because mothers and fetuses share, on average, only about 50 percent of their genes, it is not surprising that just as skin grafts often fail, the fetal graft sometimes fails, leading to spontaneous abortion or miscarriage. This is an immune challenge that is faced by no other organ in the body but it is usually mediated adequately by the placenta.[16] What keeps the maternal system from rejecting all fetuses carrying genes and their proteins that would usually be recognized as foreign to the mother's immune system? As noted, just as in the luteal phase of the menstrual cycle, the mother's immune system is depressed early in pregnancy (under the influence of progesterone) so it is not as likely to reject the fetal graft as it might at other times. This phenomenon has been referred to as the "immunological inertia of viviparity,"[17] and it is probably important for all animals that gestate internally and give birth to live young.

Maternal-Fetal Incompatibilities: The MHC Gene Complex

In some cases the mother may reject a fetus that is too much like her. This is seen in the case of immune system genes that are part of the *major histocompatibility complex*, known as MHC. The MHC is found in all vertebrates and it is extraordinarily diverse, even in humans, who have hundreds of alleles.[18] This diversity partly reflects direct effects of pathogens on the human genome throughout history, but some of it may be influenced by mating patterns that increase the likelihood that offspring will have disease resistance. (In this case, selection favors greater differences between parents in MHC genes and offspring who are genetically different from the mother.) This phenomenon is referred to as "MHC-mediated mate choice" and it appears to be mediated by olfactory cues.[19] This has been amply demonstrated in rodents, but, as noted with regard to menstrual synchrony, the role of olfaction in human behavior is still uncertain and highly controversial. In any event, even if olfaction is not influencing mate choice in humans, parental gene-sharing at the MHC locus is strongly linked to implantation failure and spontaneous abortion. In a study of a Hutterite population, couples who were similar in MHC genes took more than 2.5 times longer to achieve pregnancy and had far more pregnancy losses than couples who were genetically different.[20]

If a woman mates with a man whose histocompatibility genes are similar to hers, the resulting embryo will also be genetically similar to her. In this situation, she may not recognize the embryo when it begins to implant and may not depress her immune system to prevent rejection. One suggestion is that this is an "anti-inbreeding" mechanism and it also may explain why women who have trouble conceiving with one man are easily able to get pregnant when they have a different partner.[21] The mechanism may be particularly important in small populations where genetic diversity would otherwise be low, a phenomenon that probably characterized most of human evolutionary history.

Maternal-Fetal Incompatibilities: Blood Type Genes

Probably more common are situations in which rejection occurs because the fetus is very different from the mother. But even if the pregnancy is maintained, maternal-fetal differences may cause complications such as that seen when the mother is Rh negative and the fetus is Rh positive. In this case the fetus has antigens that are different from the mother's and antibody production is triggered by her immune system, usually at the time of delivery. Unless these antibodies are neutralized within a few hours of delivery, they will remain in her system resulting in more severe immunological problems in subsequent incompatible pregnancies. The abortion and stillbirth rate is higher in these incompatible pregnancies, as is hemolytic disease of the newborn.

ABO incompatibility is more common, although we do not hear much about it because it appears that many of the incompatible pregnancies may be lost before the mother even knows she is pregnant, perhaps before implantation. These occur most commonly when the mother is blood type O with a fetus of blood type A or B (also possible with mother type A and fetus type B and vice versa).[22] In this situation, the type O mother has antibodies against A and B antigens that circulate in her system at all times, even if there is not an immunological challenge. Thus, it is theoretically possible for these antibodies to cross the placenta and destroy the red blood cells of fetuses of type A or B, resulting in a number of complications and even miscarriage. Clinically recognized problems are not as common as would be expected, however, and although there are cases of elevated red blood cell counts in newborns of ABO-incompatible pregnancies,[23] this does not appear to affect the infant's long-term health.

As noted, one reason that ABO-incompatible pregnancies do not come to the attention of clinicians is that most are probably rejected before or soon after conception. In an early survey covering the years 1927–1944, there was a net deficiency of 25% of A children in father A–mother O matings compared to the opposite configuration (father O–mother A). This meant a fetal death rate of 8% of all A children or 3% of all conceptions for the population they studied. These figures are much higher than the contemporary Rh incompatibility death rate of 0.5%. These researchers argued that the reason that ABO hemolytic disease of the newborn is so rare is that most of the pregnancies that would lead to the disease are aborted or miscarried before the disease is manifested.[24] Other studies confirmed the lower fertility in couples who were incompatible in genes of the ABO complex.[25]

Although these findings have been controversial, it is clear that the placenta is not a perfect barrier against fetal antigens that may trigger the mother's immune response, just as it is not a very good barrier against noxious agents that may affect the fetus. Is there evidence that the human placenta with a thinner barrier is less protective than other types with thicker layers? A particularly tragic example of the problems that arise when drugs pass to the fetus through the placenta was the use of thalidomide in the 1950s and 1960s as a drug to suppress nausea of pregnancy. When the drug was tested on galagos, primates with thick-barrier placentas, it had no apparent effect on the developing fetus.[26] But when the drug was used by humans, serious limb malformations occurred in the fetus, crippling thousands of European children (fortunately, due to a particularly vigilant member of the U.S. FDA, Frances Oldham Kelsey, the drug was prohibited in the United States). One of the problems with testing drugs for use in pregnancy is that placenta types are so variable across species, even within taxonomic groups, that tests done on other species (like macaques and rats) may not be meaningful for understanding human pregnancies. Of course, it is obvious that testing drugs directly on pregnant women would not be ethically or morally acceptable, despite the potential benefits.

Early Pregnancy Loss and Maternal-Fetal Conflict

Evolutionary biologists often talk about pregnancy as a time for potential mother-fetal conflict because the interests of the mother and fetus do not always coincide.[27] If a pregnancy interferes with her own health or the health of her current and future offspring, it may be in the mother's best interest to abort, but it is clearly in the interest of the fetus to maintain the pregnancy, no matter how bad the situation is.

Based on studies in health-rich populations, it has been estimated that more than half of all conceptions are lost within the first five to six weeks, most of those occurring before implantation[28] (see Figure 3-1). Perhaps only a third of conceptions result in a healthy, full-term infant.[29] Preliminary studies of women in health-poor populations suggest that the failure rate is even higher.[30] If reproductive success is what it is all about, then how can there be benefits to early pregnancy loss? One way to examine this is to recall that for humans, quality of offspring overrides quantity because of the huge investment made by women and couples in each pregnancy and the subsequent years of parenting. An evolutionary perspective argues that gestating, giving birth to, breastfeeding, and raising an offspring that is not healthy and capable of reproducing would be energy that could be better allocated to future offspring who are healthier and more likely to reproduce. Indeed, in one study 77% of spontaneous abortions in the first trimester had chromosomal abnormalities,[31] although the portion of all lost conceptions due to genetic abnormalities is probably closer to 15%.[32] From the medical perspective, however, reproductive failure is often seen as pathological or at least unhealthy and preventing it is an overriding goal for physicians, women, and couples, serving as an example of where medicine and evolution might be at odds.[33]

Certainly there are examples of early pregnancy failure in which there are clear clinical explanations such as those caused by sexually transmitted diseases, other infectious agents, or uterine and placental abnormalities, but in many cases there are less obvious explanations, which may be effectively sought in evolutionary understanding of strategies for reproductive success. This might be seen as a "cut your losses" strategy whereby a pregnancy is terminated early if the fetus is not likely to be viable or healthy or otherwise jeopardizes future reproductive opportunities. As Vitzthum notes, in many cases of early

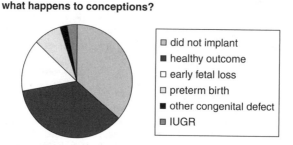

Figure 3-1 Fate of conceptions. From Power and Tardif, 2005, page 88.

pregnancy loss, "the woman is not necessarily broke, and there may not be any reason to fix her."[34]

Timing in terms of resource availability may also play a role in early pregnancy loss. Vitzthum found that when Bolivian women engaged in particularly strenuous labor, such as during the planting and harvesting seasons, their rate of pregnancy loss was almost four times the risk during other seasons.[35] In general, women who work in the agricultural sector have higher losses than those who are not engaged in hard physical labor. Psychosocial stress may also have an impact on pregnancy loss,[36] most likely through elevations of the hormone cortisol.[37] This adds support to the evolutionary view that successfully bearing and raising offspring in humans benefits from social support to reduce stress at all stages.

The risk of losing a pregnancy also rises with age, most likely associated with declining maternal health, and it is especially pronounced in women from health-poor populations.[38] If health status were equal, however, an evolutionary perspective would predict that older women would have a higher threshold for rejecting fetuses that were less-than-optimal if their future chances of reproducing were limited by age. In other words, the "cut your losses" concept would no longer work and would be replaced by "anything is better than nothing." Perhaps in support of this, children with congenital defects are more common in births to older women. In fact, the rate of spontaneous abortion of fetuses with chromosomal abnormalities is much higher in women aged 25–29 than in those aged 30–39.[39]

Although the termination rate declines after the first trimester,[40] the potential for maternal-fetal conflict continues, especially if there is nutritional competition. The timing of nutritional stress may be important. Primate researcher Suzette Tardif and her colleagues demonstrated that if marmosets experience even modest food restriction in early to mid pregnancy, fetal loss frequently occurs.[41] The same degree of food restriction later in pregnancy, however, does not usually result in loss, although preterm deliveries are more common. From an evolutionary perspective, fetal loss in early pregnancy under conditions in which the infant may have compromised health makes sense in that it occurs before significant investment in the pregnancy has been made by the mother. Later, however, the mother's reproductive success may be better served by maintaining the pregnancy until birth, at which time breastfeeding and other forms of maternal care may be able to make up for what was lost.

In addition to pregnancy having the potential for mother-infant conflict, it is also a time of father-mother conflict in evolutionary terms. A particular pregnancy may be the only one the father will have with a given woman, so it may not be important to him (in an evolutionary sense) to compromise a current pregnancy for a better return in a later pregnancy. For the mother, however, she may be able to have more pregnancies, so if the current one seems compromised, terminating it and starting again later may be a reasonable strategy for her.

Gestational Diabetes

Other clinical conditions that benefit from consideration from an evolutionary perspective and that may also result from maternal-fetal competition for nutrients include gestational diabetes and eclampsia/preeclampsia. In these cases, changes in maternal physiology that work to increase delivery of oxygen and nutrients to the fetus have passed from "normal" to problematic and usually warrant medical treatment. Furthermore, women who develop gestational diabetes are at higher risk for type 2 diabetes later in life.[42]

For women without diabetes, blood glucose levels rise after a meal and return to lower levels when counteracted by insulin. During pregnancy, however, both glucose and insulin levels remain elevated for a longer time to benefit the fetus. In evolutionary biologist David Haig's view, this is an example of the fetal interests working against the maternal interests as the fetus obtains more and more glucose at the expense of the mother's health. Because glucose elevation is also associated with excess nutrient intake in pregnancy, it is more common in health-rich countries. Women who develop gestational diabetes often give birth to large infants who are themselves predisposed to developing diabetes later in life.[43] This suggests that there is likely an optimal level of nutritional intake during pregnancy and that both reduction and excess of nutrients can cause pregnancy complications and lifelong health problems for both mother and infant. Thus we have another example of balancing selection, whereby both ends of the pregnancy food-intake scale result in reduced reproductive success.

Eclampsia and Preeclampsia

Preeclampsia and its more severe form, eclampsia, is a complication of pregnancy worldwide, found in both health-rich and health-poor populations. Preeclampsia is associated with hypertension in the mother and is estimated to occur in as many as 10% of births. The only "cure" for preeclampsia is delivery of the fetus and placenta, so it is a common reason for premature birth. If the pregnancy is not terminated, the mother will experience damage to the kidney, liver, and brain, and maternal convulsions may result if it proceeds to eclampsia. Furthermore, even though the short-term effects may be improved with delivery, there is evidence that there may be lifelong effects on maternal health.[44] Unfortunately, there is no animal model for preeclampsia, which has inhibited research on causation and potential interventions. Some scholars[45] argue that the disorder is related to increased brain size in *Homo sapiens* and suggest that it may be uniquely found in our species.

One notable effect of increased brain size in our species is increased need for oxygen and nutrients, as discussed earlier. At the time of implantation of the embryo, the trophoblast is imbedded in the uterine wall in such a way that it co-opts maternal blood vessels to provide for the needs of the developing embryo. This process of implantation of the trophoblast within a few days of conception is pretty much the same in all mammals. In humans, however, there is a second co-option of maternal blood vessels (known as a secondary invasion of the trophoblast) that occurs in the third month of gestation when brain development in the fetus takes off and nutrient and oxygen needs increase significantly. Sometimes this second invasion is incomplete or fails altogether. Among the reasons for this may be maternal-fetal similarities in MHC genes[46] or the result of maternal-fetal conflict for nutrients that the infant "wins."[47] In this latter case, preeclampsia "makes sense" from an evolutionary perspective because it serves as a way for the fetus to divert resources from the mother to itself.

In response to increasing nutrient and oxygen needs of the fetus and the limited vascular support provided by the incomplete invasion, the mother's own blood flow system must compensate, with the resulting elevation of blood pressure. Slight elevations of blood pressure are common in pregnancy, but when it becomes excessive, the disorder of preeclampsia is diagnosed and women who are receiving prenatal care are monitored closely and usually delivered early. Where prenatal care is absent, death of the mother and infant often results: the World Health Organization estimates that as many as 70,000 women die each year from eclampsia/preeclampsia. This suggests that eclampsia/preeclampsia was likely a major agent of natural selection in the past when early induction of labor and delivery of the fetus was not an option.

Noting that preeclampsia is usually restricted to first pregnancies, immunologist P. Y. Robillard and his colleagues suggest ways of reducing the incidence of preeclampsia through understanding the evolutionary background and by relating the disorders to other aspects of human reproductive ecology including highly frequent and nonovulatory sexual activity. There are other interpretations and recommendations for treatment of preeclampsia,[48] but an evolutionary perspective offered by Robillard and his colleagues has a somewhat simple recommendation: delay pregnancy until after several months of sexual activity, preventing what they refer to as not just a disease of first pregnancy, but a "couple disease" resulting from exposure to novel sperm.[49] They argue that delaying pregnancy for a few months gives the woman's immune system time to adjust to her partner's antigens, decreasing the likelihood that her system will challenge the fetal "allograft" in the early months of pregnancy.

So if a woman wants to try to prevent developing preeclampsia, Robillard and his colleagues would advise her to wait a while before attempting pregnancy with her partner. They may also suggest that she have sex with this partner frequently so that her system can adapt to his sperm and the associated antigens. In this rather speculative view, highly frequent sexual activity in human partnerships may be protective against the failure of the second deep invasion of the trophoblast. From an evolutionary perspective, women in the past who engaged in frequent nonovulatory sexual activity may have had more successful pregnancies than those who engaged in sex only when they were ovulating.

Unfortunately, the evolutionary view of pregnancy is one of stress and conflict, rather than the unrestrained joy that many women feel when they first learn that they have conceived. Of course, the joy is understandable and appropriate in a wanted pregnancy, but the first few months are often not much fun for the mother because it is the time of morning sickness and physiological changes that may be strange and disconcerting. Once the pregnancy "takes," however, it takes off and usually results in a healthy offspring. But as we will see in the next chapter, getting pregnant is not the whole story. We begin with a bit more commentary on the first trimester and then discuss the importance of nutrition throughout pregnancy, continuing the refrain that reproductive success is about food.

4

Staying Pregnant

As we have seen, getting pregnant is not as easy as it might sometimes seem. Natural selection has played an important role in mammalian evolution by ensuring that conceptions that are not likely to survive pregnancy and the early months of life are eliminated before the mother has invested much time and energy. Today in health-rich nations many conceptions that would have been lost early in pregnancy or in the first few days of life are "rescued" by technological interventions that are usually welcomed by women and men who desire children, no matter what their reproductive or evolutionary "value" might be. In fact, many readers of this book might not be here today if they had been conceived, gestated, or born under conditions that existed a mere 100 years ago, much less earlier in human evolution. Although getting pregnant and staying pregnant are very different challenges, thankfully, once the first two to three months of pregnancy have passed, most proceed with little stress. Before moving on to the last two trimesters, a few concerns with regard to the first trimester should be mentioned, including morning sickness.

More on the First Trimester

During the first trimester, the developing organism is referred to as an embryo. This is the period of differentiation during which most of the organ systems are being developed. It is also a period of great vulnerability because many factors can interfere with the developing embryo to render it nonviable or to affect development in a way that will have lifelong effects if the embryo survives. Here are some of the systems that develop in the first eight weeks of pregnancy: circulatory system (week 2); nervous system (week 3); limb buds, heart, and most organs (week 4). Brain and sexual development accelerate during week 5, and by week 6, the testes are obvious and can be seen on ultrasound.

As noted, the first trimester is a period of great vulnerability and it is often a time when a woman does not even realize she is pregnant. This is particularly problematic when her normal diet is not sufficient or well balanced, a common problem with teen pregnancies in the United States. Nutritional deficiencies at this time can result in numerous developmental failures in the embryo: insufficient riboflavin can cause problems with skeletal development; low B-6 (pyridoxine) can result in neuromotor problems; low B-12 can result in hydrocephaly; low niacin can result in cleft palate; low folic acid can cause neural tube defects; low vitamin A can cause vision problems; and low iodine can result in neurological problems and cretinism.

Iodine Deficiency, Cretinism, and the Evolution of PTC-Tasting

Cretinism, a condition in which individuals show marked physical and mental abnormalities, results from an underactive maternal thyroid during pregnancy. This can be due to a clinical condition in the fetus or the mother or it may be due to insufficient intake of dietary iodine in pregnancy. In both cases, if the infant is treated soon after birth, the condition does not appear to affect later physical development, but if it is not treated, the child will be stunted and, perhaps, mentally retarded. Iodine deficiency during critical periods of brain development may not be reversible.

Dietary iodine is used in the formation of thyroid hormone. Cretinism caused by insufficient iodine in the diet is a worldwide problem and is most common in parts of the world where crops are grown in iodine-poor soils. The World Health Organization estimates that approximately 35% of the world population is at risk for iodine deficiency ranging from a low of about 10% in the Americas to a high of about 57% in Europe.[1] Iodine-deficient soils are most commonly found in areas that were previously under continental glaciers, which explains the high frequency of low-iodine intake in Europe before fortification of salt and other foods. A deficiency of this mineral also causes goiter in adults, and this region is often referred to as the "goiter belt" because of high incidence of the disorder. Goiters are signs of an overactive thyroid gland, which expands to form a mass in the throat area and is very visible in extreme cases (Figure 4-1). In most cases the goiter does not have a major impact on the health of the mother, but it has an obvious and dangerous effect on her developing fetus and, thus, her reproductive success. Dietary iodine deficiency is the primary reason for the iodization of salt and other foods, so there is a technological fix, but there is speculation that there may also be a genetic "fix" that results from natural selection.

An interesting, although still speculative, proposal is that the high frequency of a common gene that affects whether a person can taste certain bitter substances may be due to its role in prevention of goiter and cretinism. This gene is commonly called the "tasting or nontasting" gene because it determines whether a person can taste the chemical known as PTC (phenylthiocarbamide). There is a great deal of variation in the frequencies of the alleles, indicating that natural selection acted on the gene in the past and may be acting on it today.[2] On the surface, it is hard to

Figure 4-1 Woman with a goiter. Her child may have suffered from malnutrition in utero or infancy. Photo by Kathy Dettwyler. Reprinted by permission of Waveland Press. From Dettwyler, *Dancing Skeletons* (Long Grove, IL: Waveland Press, 1994). All rights reserved.

imagine how a person who can taste the substance would be advantaged over the nontaster or vice versa.

The chemical is closely related to those that cause bitter tastes in members of the cabbage family such as broccoli and brussels sprouts. Tasters (those with the homozygous dominant or heterozygous combination of alleles) detect the bitter substance and tend to avoid eating such foods or eat them in small quantities. Nontasters (those with two recessive alleles) are not as aware of the bitter taste and tend to eat the foods in larger quantities. Interestingly, these foods are known as "goitrogens" in that they inhibit the uptake of iodine, which is a problem in populations that consume foods grown on iodine-deficient soils. Before the iodization of salt, women who avoided eating these bitter foods (because they were tasters) would have been better able to utilize the iodine available and less likely to have children with cretinism. Nontasters may have been more likely to eat the foods, and thus, their iodine uptake may have been compromised, resulting perhaps in major health problems in their children. So if brussels sprouts are not among your favorite foods, you may be able to thank your aversion to their bitter taste for keeping your ancestors healthy. Furthermore, if cabbage-family vegetables were particularly likely to cause morning sickness in your mother (as will be discussed next), you may have been protected from developing thyroid problems or even cretinism. Cabbage-family vegetables are among those most frequently cited as causing morning sickness.

Medical Concerns (or Not) of Early Pregnancy:
Morning Sickness

In the previous chapter we discussed two common medical concerns of pregnancy that have been viewed as nonpathological in some circumstances by evolutionary medicine: early fetal loss and eclampsia and preeclampsia. A third example is nausea of pregnancy. Although morning sickness (also called "nausea and vomiting of pregnancy" or NVP and early pregnancy sickness) is so common (its incidence in the United States may be as high as 90% of pregnancies) that it is a normal and expected part of pregnancy, there are still efforts to "treat" it by developing drugs or other interventions that can prevent its occurrence. As noted in the previous chapter, a particularly tragic effort to "solve" the problem of morning sickness in the 1950s with the drug thalidomide resulted in thousands of children being born with severe abnormalities. Morning sickness is an example of a health problem that benefits from questioning whether it is a defense or a defect.

Evolutionary biologist Margie Profet[3] proposed that nausea during early pregnancy evolved as a protection against toxins and other dangerous substances that could harm the developing embryo and thus may be a defense rather than a defect. The direct or proximate cause of nausea and food aversions is likely hormonal (perhaps HCG, progesterone, and estradiol, which rise early in pregnancy and changes in gut hormones that delay gastric emptying), but an evolutionary or ultimate explanation is that pregnant women who found potentially harmful food components to be aversive may have protected their fetuses from developmental damage, especially during the first trimester. If this is true, one could speculate that ancestral women who had morning sickness had more healthy offspring and greater reproductive success. There is extensive evidence that pregnancies that do not include morning sickness may be at risk. Miscarriage is far less frequent in women who report than in those who do not report feeling nauseous early in pregnancy.[4] Midwives with whom I have worked are concerned when they have a client who reports not having had morning sickness.

Foods to which women report aversions include pungent meats, bitter vegetables (such as the brussels sprouts discussed earlier), overripe foods, spicy foods, and smoked foods, all of which may have compounds that interfere with fetal development. It is important to note that in most cases these food aversions developed during pregnancy and most women report that they previously liked the foods. In an extensive survey of the literature,[5] the category "meat, fish, poultry, and eggs" was listed most frequently as food aversions, which had not been expected based on previous research and the preconceived notion that all of these foods would be "good for" pregnant women to eat. If we look at the conditions under which meats and poultry would have been stored in the past and in many areas of the world today it becomes clear that food-borne illnesses and food poisoning are much more common for this category of foods than for fruits, vegetables, and beverages. Exposure to these kinds of contaminants is especially problematic for women during the early stages of pregnancy when their immune systems are suppressed.[6] Later, when the fetus is not so vulnerable, a mother's aversions to meats, eggs, and poultry decrease when the benefits of eating these protein-rich foods outweigh the risks to the developing baby.

The weeks when morning sickness is most pronounced coincide with the time when the embryo is most vulnerable. Although there is concern that chronic nausea can lead to mal- and undernutrition, most problems with morning sickness are during early pregnancy when lowered food intake by the mother is not as problematic as it would be later in pregnancy when rapid fetal growth is occurring. In most cases, nausea disappears after the first trimester. The trade-off here is that mothers may be miserable for a period during pregnancy, but the fetus, who is the cause of the misery, is protecting itself.

Cultural taboos are common for women during pregnancy and include prohibitions against certain behaviors and consumption of certain foods. Many anthropologists have interpreted the taboos as cultural mechanisms for protecting pregnant women at a vulnerable time. Anthropologist Dan Fessler,[7] in an extensive review of the cross-cultural literature, found 73 societies that had specific food taboos for pregnant women. Meat was by far the most common category of forbidden food. Even handling of meat is often prohibited for pregnant women and, as we know, handling can expose a person to dangerous contaminants. Given the importance of meat consumption in human evolution,[8] Fessler concludes that "meat-borne diseases have constituted a significant source of selective pressure on pregnant women for much of human history."[9]

One problem that pregnant women may encounter when meat is not a prominent part of their diet is iron depletion. Iron needs increase in pregnancy and iron deficiency is the most common nutritional deficiency in the world today. Insufficient iron is implicated in preterm birth, low birth weight, infant and maternal mortality, and delayed cognitive development of the child. Fessler argues that iron depletion early in pregnancy could be adaptive in that low iron levels inhibit pathogen proliferation, which may compensate for the mother's suppressed immune system at this time. One function of geophagy (see p. 80) may be that it serves to inhibit the absorption of iron. As pregnancy proceeds and the iron needs of the fetus increase, the mother's physiology changes so that she can better absorb what iron is available. Trade-offs would have to be examined—are babies healthier if their mothers risked low iron to suppress infection early in pregnancy or are the dangers of low iron greater than the dangers of infection?

As far as we know, morning sickness does not occur in other animal species, although decreased appetite early in pregnancy has been reported for domestic dogs, captive rhesus macaques, and captive chimpanzees.[10] Cross-cultural studies of morning sickness confirm that it is not restricted to women in health-rich countries, although it seems to be somewhat more rare where traditional foods are bland (such as regions where corn or rice is the primary staple). Morning sickness was probably more important in early human evolution when foods were less predictable and more variable; this degree of variability has come back into human diets only with mass marketing of foods.

We can argue that morning sickness, up to a point, can be a good thing, but there is a point when the defense becomes a defect. There are limits to the adaptive value of nausea and vomiting early in pregnancy and when nausea results in dehydration and extreme weight loss, health of both mother and fetus is compromised and one or both could die. At this point, the nausea moves into the category of a "defect" and is estimated to occur in 1 in 5,000 pregnancies.[11] Ewald proposes that the extreme forms of pregnancy sickness may result from a naturally selected defense being "hijacked" by an infectious agent.[12] Anthropologist Ivy Pike argues that the "embryo protection hypothesis" for morning sickness has been largely based on observations of well-nourished women and contends that for women who were inadequately nourished before pregnancy, there are nutritional consequences. She presents data for African Turkana women showing that the risk for fetal, perinatal, or neonatal mortality is more than twice as high if the woman reported morning sickness.[13]

Another hypothesis for nausea of pregnancy is that it is a result of maternal-fetal conflict early in pregnancy, particularly in cases where the resulting nausea is so severe that maternal health is compromised or under conditions of chronic nutritional stress. In this view, morning sickness is a by-product of maternal-fetal competition for nutrients and is not itself a product of selection.[14] For the mother, the worst that she can do (from a fitness standpoint) is continue a pregnancy that will result in a nonviable or otherwise compromised infant. Thus, it may be important for her embryo to signal that it is healthy and likely to develop optimally. Given that lack of morning sickness is associated with miscarriages, it may be that nausea in early pregnancy is a signal of embryo viability[15] and thus has selective value in itself.

A phenomenon somewhat related to food aversion is the craving for clays (geophagy) reported by pregnant women in a number of societies. In medical literature, this craving is often reported as pathological, but its existence is so widespread that scholars of evolutionary medicine search for adaptive explanations.[16] Worldwide, clays are used to relieve diarrhea (the original Kaopectate, after all, was mostly kaolin, a clay), detoxify compounds, and provide minerals that are insufficient in the diet. In Africa, this practice is also employed by women seeking to relieve nausea of pregnancy and it can serve to bind toxins that would harm the fetus at this stage.

When geophagy continues beyond the early stages of pregnancy, it probably adds important nutrients to the diet, especially calcium, essential for fetal skeletal development and maintenance of blood pressure in pregnancy. Anthropologists Andrea Wiley and Sol Katz propose that clay as a source of calcium helps to explain the distribution of geophagy in African populations. Their survey of 60 societies confirms that where dairying is practiced and calcium is available in the diets of pregnant women, geophagy is less common than where dairy foods are not available. This work confirms that what may be seen as pathological or abnormal by clinicians may make sense from an evolutionary perspective because in many cases geophagy serves to reduce the negative aspects of morning sickness, detoxify agents early in pregnancy, and provide calcium and other minerals. Given evidence that clay consumption occurs in chimpanzees and may have been practiced by early hominins,[17] it seems to have a long evolutionary history. Certainly most of the sources of calcium sought by pregnant women today (dairy products) were not available to our ancestors and clay consumption may have made the difference between a healthy and unhealthy pregnancy.[18] As with all practices, however, geophagy may have negative consequences, including exposure to pathogens in soil, iron deficiency anemia, and lead poisoning.[19]

Second and Third Trimesters

Development of the respiratory and neurological systems continues in the second trimester, toward the end of which the fetus is said to be viable,[20] at least with modern technologies. Pregnant women will notice a lot of fetal movement early in the second trimester, beginning with somewhat frequent position changes (an average of 10 per hour) and then smoothing out to resemble what has been called "a young female Balinese temple dancer."[21] Other movements reported include sucking and swallowing, head movements, and hand-to-face contacts. The fetus even does somersaults and loop-de-loops during the second trimester, some of which account for the umbilical cord being wrapped around the neck at birth, as I will discuss later. Movements themselves and their duration increase so that some women report great difficulty sleeping. Then the movements begin to decrease as the fetus becomes larger and fills the uterus so that movement is impeded. Many of the movements exhibited by the fetus serve to prepare it for actions and reflexes that it will use after birth.

Growth in the Third Trimester

Further development in the respiratory and neurological systems continues in the third trimester and the circulatory and respiratory systems prepare for the changes that will take place at birth. The most important development that is happening at this time, however, is increase in body weight and fat deposition. Fat is particularly important for fetal development and human babies are notably fatter than the infants of other mammals[22] (averaging about 16% fatter), even under conditions of nutritional stress or intrauterine growth retardation (IUGR). The primary selective value of fat accumulation in the last weeks of pregnancy is its contribution to early postnatal survival, especially in helping to maintain that expensive, rapidly growing brain. Fat babies may also serve as a signal to a mother that the baby is healthy and "worth saving" under conditions where infanticide or neglect are options.[23]

The potential for maternal-fetal conflict continues in the last trimester, especially in competition for nutrients. It seems that severe under- or malnutrition during pregnancy would compromise the uterine environment for the fetus so that it miscarries. But, surprisingly, the prediction that near-starvation in the mother would terminate a pregnancy is not often borne out: women continue to give birth even under severe food restrictions such as occur with famine and war. This suggests that the fetus has ways of prolonging the pregnancy, even if there are negative consequences for maternal health. But just because a fetus makes it through pregnancy to be born does not mean that health is not compromised by in-utero nutritional stress. Such babies are often born small for their gestational age due to intrauterine growth retardation, and they are at much higher risk for health challenges and even death in the first few months after birth.

Some instances of IUGR result from preeclampsia in the mother, genetic inabilities to metabolize nutrients, or placental problems leading to inefficient delivery of nutrients to the fetus, even when the mother is well nourished. These causes fall in the realm of clinical concern and do not necessarily benefit from an evolutionary consideration. But IUGR caused by socioeconomic-induced under- and malnutrition can benefit from analysis by evolutionary medicine.

There is increasing evidence that growth during pregnancy can affect a person's health several decades later.[24] Table 4-1 summarizes some of the adult health problems that seem to be preceded by IUGR and its indicator, low birth weight. It is important to realize that it is not low birth weight per se that has later health consequences but that low birth weight is proxy for all of the other things that a

Table 4-1 Later-life conditions that seem to be associated with low birth weight

Adult conditions associated with low birth weight (<5.5 lb or 2,500 g) (proxy for IUGR)	Source
Hypertension	Barker et al, 1990
Cardiovascular disease	Barker et al., 1989; Barker, 1995, 1997
Diabetes type 2	Barker, 1999, 2005
Early sexual maturation/menarche	Gluckman, Hanson, and Beedle, 2007
Depression in girls	Costello et al., 2007
Obesity	Gluckman, Hanson, and Beedle, 2007
Osteoporosis	Gluckman, Hanson, and Beedle, 2007
Impaired thymus growth/impaired immunocompetence	McDade et al., 2001

fetus experiences while gestating in a nonoptimal environment. If food is restricted, for example, the hungry developing brain is served first and other developing organs get what is left over, which may not be sufficient for normal growth. Thus, a fetus that was malnourished in pregnancy will have a smaller-than-normal liver, pancreas, and gut. Compromise in the functioning of these crucial organs further compromises cholesterol and glucose metabolism, thus explaining the link between IUGR and several health challenges in adulthood. Intrauterine growth retardation that leads to impaired thymus growth means compromised immune function so that even at younger ages a person has limited ability to fight off infections.[25] Additionally, it is also important to remember that early postnatal environments have an impact on later life health so we cannot just focus on pregnancy.[26]

Some scholars working in this field use the hypothesis of "fetal programming" to explain the relationship to adult health. Consider that a computer program will work fine if it is written correctly but will malfunction if some of the code is wrong. In the same way, if the fetal "program" develops correctly in a healthy pregnancy, later life health and development have a better chance of proceeding healthfully than if the program has errors because of malnutrition or other insults during pregnancy. Of course in some cases, there may be genetic factors that affect both birth weight and adult diseases, but in other cases, it may be that the program is altered by environmental circumstances.

The fetal programming hypothesis proposes that a developing fetus uses cues to assess not only the environment of gestation but also the postnatal environment. The assumption is that if there is nutritional stress during pregnancy, there will also be nutritional stress after birth. Thus, the baby's system is programmed to be ready for the same conditions it experiences in utero. The result is a metabolism that is very efficient at getting the most out of whatever calories are available—a thrifty metabolism. When this get-the-most-out-of-every-calorie metabolism meets food abundance in childhood, excess fat deposition and other health challenges may result. Furthermore, women in health-rich populations who try to restrict their food intake during pregnancy to avoid gaining weight may be inadvertently teaching their fetuses to expect insufficient nutrients after birth, which can be problematic in a world of high-fat diets and more than enough food.

Some of the greatest challenges to adult health are seen in people who were born small after nutritionally stressful pregnancies but subsequently grow up in environments that are very different from the ones for which they were programmed prenatally, resulting in a "mismatch" (Figure 4-2). This is particularly true today with rapid globalization and the associated high-fat and high-caloric foods that are often readily available. The baby may be able to put on weight and reach a normal level rather quickly, but the suppressed growth of most organs of the body cannot usually be made up for postnatally. An overweight child who was low birth weight but is eating lots of fats and sugars puts more and more stress on her liver, pancreas, and gut. No wonder she finds she has problems with obesity, diabetes, hypertension, and atherosclerosis when she is older.

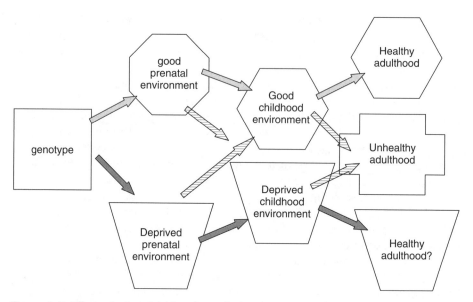

Figure 4-2 Effects of mismatched and matched environments of pregnancy and child development on adult health. Solid arrows are matched; hatched ones are mismatched. Adapted from Gluckman et al., 2007, Figure 2, page 8.

Some have argued that children born small are better off if they remain small throughout their lives and that efforts to improve health by improving nutrition may misfire.[27] In this case, if the prenatal and postnatal environments are similar, the metabolic needs developed in utero would be better matched to the world into which the child is born and grows up and it is predicted that adult health would not be further compromised. Unfortunately, such arguments often lead to policy decisions that fail to improve conditions for women during pregnancy, especially in circumstances where the causes of poor conditions are social and economic inequalities.[28] In this case, the argument that the status quo is best does not resonate morally and ethically, nor does it work in populations undergoing modernization. Additionally, nutritional stress during pregnancy can lead to later life health challenges even if the infant is not small at birth.[29] Furthermore, infants who are unusually large at birth are predisposed to some adult-onset diseases, suggesting a much broader phenomenon with regard to intrauterine developmental effects on later life health.

Some of the most astounding and worrisome findings from studies of the relationship between prenatal conditions and later life health are that the effects are transgenerational. As noted previously, the quality of the eggs developing in a girl fetus just a few weeks after fertilization will determine the quality of her offspring, several years later. Those eggs have a "memory" of their time in utero that affects their quality when they are ovulated and fertilized to start a new life trajectory. Anthropologist Chris Kuzawa[30] has proposed the "intergenerational phenotypic inertia" hypothesis, which states that the fetus is obtaining cues not just from the mother, but from her entire matrilineage. This suggests that adaptations that have been successful for generations because they served to buffer pregnancy from insult due to poor nutrition may not be amenable to short-term fixes. Because the effects are not just limited to a single pregnancy, the implication is that public health measures to improve infant birth weight should begin long before pregnancy and should not be judged as successful or failed on the basis of data from a single generation. Peter Ellison[31] has proposed that this intergenerational linkage to physiology may characterize a number of chronic diseases and disorders that seem to be related to early life events. This is further support of the argument that only by improving health for all pregnant women and their offspring can we eventually break the vicious cycle of transgenerational prenatal programming.

Pediatrician Peter Nathanielsz[32] has argued that prenatal programming can explain good adult health as well and proposes that it resolves the problem of the "French paradox" whereby the French can eat high-fat and high-cholesterol foods all of their lives without developing the cardiovascular problems and diabetes that their diet would inflict on most Americans. He notes that for more than a century, the French have had a very sophisticated system of prenatal care for all pregnant women that ensures their fetuses develop optimally. Healthy pregnancies yield healthy children yield healthy adults, no matter what the postnatal environment is, just as unhealthy pregnancies yield unhealthy adults, even in seemingly optimal postnatal environments.

Prenatal programming is affected by a number of factors other than nutrition with infection and inflammation being predominant.[33] In fact, early life infections may have as much of an impact on later health as prenatal nutrition. In other words, it is not just undernutrition alone or even infections alone that predispose to poor adult health, but the synergistic effects of these two factors. This is not surprising, given the obvious and well-known synergy between malnutrition and infection in childhood health.[34] Of course any model linking to adult disease must include genes and environment and their interaction.

The Pregnant Biped

Bipedalism is the hallmark of the human species, but it does not come without costs, especially with regard to the spine. Four-legged animals have curved spinal columns and carry their internal organs (and babies when they are pregnant) slung beneath the horizontal spinal supports. When our ancestors began to habitually walk upright, however, the spinal column assumed an S-shape in order to keep the body mass above the legs and feet that supported it. Thus, the center of gravity of a biped is just above the hip region. The curve in the lower back is known as the

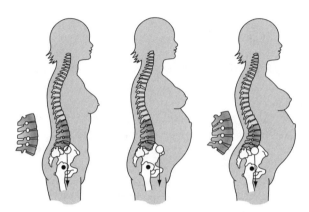

Figure 4-3 Effects of pregnancy on upright posture. From Whitcome et al. 2007, page 1075. Reprinted by permission from Macmillan Publishers Ltd: *Nature Genetics* 450, 2007.

lumbar curve and the posture assumed is called lordosis. It is a weak point of the skeleton and accounts for the almost ubiquitous lower back pain that is associated with aging.

As a woman reaches the last trimester of pregnancy, the baby she is carrying projects farther and farther in front of her, throwing off her center of gravity. Fortunately, her spine has the ability to compensate for this shift in balance, thanks to natural selection for greater wedging of her lower (lumbar) vertebrae. Anthropologist Katherine Whitcome and her colleagues studied the spines of several women late in pregnancy and found that they were able to increase the curvature of their lower spines to keep the center of gravity positioned above their hips, just as it is when they are not pregnant.[35] They could do this because more of their lumbar vertebrae were wedged toward the back in comparison to men (Figure 4-3). Furthermore, the researchers found that the australopithecines also showed this adaptation, suggesting that it traces its origin to the beginnings of bipedalism itself. Unfortunately, it also means that women have more lower back pain during pregnancy and are at higher risk for slipped disks in this region.

Psychosocial and Other Stress during Pregnancy

One mechanism that has been proposed to affect fetal programming is excess production of hormones known as glucocorticoids, the best known of which is cortisol, often produced in response to stress. Excess glucocorticoids are also implicated in a number of the adult-onset diseases that have been linked to low birth weight, so it has been suggested that this excess can explain both sets of phenomena.[36] One proposal is that excess glucocorticoids, especially late in pregnancy, can alter the development of several systems, including metabolic functions; insulin resistance; cardiovascular, liver, and pancreatic functions. Stress is also known to compromise immune function.[37] Glucocorticoids are also important for brain development, and in excess, can cause problems, particularly if the excess occurs during sensitive periods. Thus, the timing of a stressful incident has varying impacts on the effects that glucocorticoids have on developing fetal systems. Attention to some of these effects is important because glucocorticoids are commonly used in medical treatment of pregnancies at risk for premature delivery.

Not surprisingly, the effects of stress hormones on developing systems are also manifested in physical and mental problems later in life. For example, stress during pregnancy appears to put children at risk for behavioral disorders, including hyperactivity, impaired cognitive function, anxiety, and fearfulness.[38] As long as the fetus is gestating, it is subjected to the physiological effects of stress that the mother is experiencing. The effects are less direct once the baby is born, and an important advantage gained by giving birth while fetal brain development is still going on is that it minimizes the direct effects of stress on the mother during later phases of neurological development.[39]

Just as a fetal environment that suffers from under- and malnutrition programs the fetus to expect similar challenges after birth, one that includes a great deal of stress on the mother may program the fetus to expect a lot of stress later in life. The result tends to be an over-active and over-reactive stress response that can itself affect later life health, both physical and mental. Cardiovascular health is worse and depression is more frequent in people who experience a great deal of stress (and the associated glucocorticoids) in utero. This is another experience that may transcend generations. If a woman who experienced stress in utero has an over-active stress response she may have elevated stress levels when she becomes pregnant and can pass this along to her own children. Those children who are girls can continue this process on down through the generations. Following a "prenatal prescription" for minimizing stress may be the only way out of this cycle.[40]

Psychosocial stress has certainly been implicated in pregnancy complications, but there is evidence that environmental stresses such as those caused by earthquakes can also affect pregnancies, depending on when they occurred. These are the sorts of stresses our ancestors may have faced—the stress of an unknown shaking of the earth, eruption of a volcano, thundering of wildebeests, screams of lions and leopards. Evidence in support of this view was found in a study of women who experienced the 1994 Northridge, California, earthquake during their pregnancies. Women who were in the first trimester when the earthquake occurred gave birth significantly earlier than those who were in the last trimester.[41]

In the ancestral past a healthy and active stress response was probably a good thing for pregnant women who had to move quickly to get out of the way of a predator or other "real" threat. The elevated glucocorticoids did their job to get our ancestors moving but then dropped off, relaxing the stress response. Today, that same healthy stress response that may have enabled our ancestors to survive is activated several times a day by bad news on TV, sirens and other loud noises, not enough money, slights or insults from other people, anger at our spouses or parents, unruly children, and our active imaginations. These problems with the stress response are by-products of modern lives and have taken a formerly advantageous response and turned it into one that causes mental and physical health problems throughout life.

Another type of stress actually appears to enhance maternal and fetal health, and that is the stress associated with moderate physical activity. From an evolutionary perspective, the idea that pregnant women should sit quietly still and rest through the last part of pregnancy does not make sense. There is no evidence from contemporary or recent past populations of people who live like our ancestors that women altered their activities in significant ways until the very end of pregnancy. Thus, it is not surprising that babies of mothers who exercised during pregnancy are easier to calm (their stress response is not over-active), seem more aware of their environment (and thus learn more), and seem to have an easier time with birth. The mothers themselves report feeling better and having shorter labors and fewer complications at birth.[42] Of course, women with other health complications would not necessarily be advised to exercise in pregnancy, and it is probably not a good idea for a woman to *start* exercising vigorously if that has not been a part of her normal lifestyle.

Finally are the stresses of toxic and other noxious substances that the mother and her developing fetus are exposed to in their everyday lives.[43] These stressors include cigarette smoke (including passive smoke), pollutants in water and air, recreational and therapeutic drugs, alcohol, caffeine, and household chemicals. It has been suggested that some cases of attention-deficit hyperactivity disorder (ADHD) can be traceable to exposure to environmental hazards. Smoking during pregnancy has been linked to low birth weight in the baby, smaller placentas, higher incidence of respiratory diseases in childhood, and higher likelihood of smoking as an adult (another form of fetal programming?). Excessive alcohol consumption during pregnancy can lead to fetal alcohol syndrome (FAS), a condition in which a baby is born with physical and mental abnormalities that are irreversible. Slowed postnatal growth can also result from alcohol use during pregnancy, the probable mechanism for which is a decreased number of cells because of alcohol's inhibition of cell division early in development. Because fermented beverages were rare to nonexistent during early human evolution, this is another mismatch between pregnancies that evolved not to expect alcohol consumption and our contemporary lives.

Clearly, any way that one can reduce stress and exposure to stressors during pregnancy is likely to be beneficial. Considering the dense social networks in which our ancestors experienced pregnancy, it is not surprising that social support enhances the health of both mother and fetus. As we will see in the next chapter, having social support at the time of birth may have made the difference between life and death for many mothers and babies in the past. This desire for the presence of others during labor and delivery and in the early months of life of our infants may be a legacy we have inherited from our ancestors.

When I began this chapter I noted that it would probably emphasize all the things that can go wrong in a pregnancy and, indeed, this seems to be just what I have done. But pregnancies usually proceed without mishap in environments of adequate food, few infectious agents, and relatively low stress. Most of us are here today because our mothers did not lose us early in pregnancy, got sufficient nutrients including vitamins and minerals, and were able to avoid serious morning sickness, gestational diabetes, and eclampsia. For those of us whose mothers had clinical problems during pregnancy, they likely had access to modern medical resources that enabled them to continue the pregnancy and give birth to a healthy infant. As I asked at the beginning of this chapter, how many of us would be here today if our mothers had been pregnant and given birth under conditions faced by our ancestors several hundreds and thousands of years ago?

5

Welcome to the World

Although the emphasis of much of the preceding chapter seemed to be on what can go wrong, most pregnancies that survive the first few months make it to birth. But it is at the time of birth that we see evidence of some of the most significant trade-offs resulting from the human evolutionary process. The most obvious one is fetal head size versus mother's pelvis. Put very simply, the relatively narrow pelvis that is good for efficient bipedalism is not very good for babies to pass through because of their large heads. In many instances, they can only be delivered with cultural or technological assistance.

Bipedalism and Birth

Perhaps no other aspect of human evolutionary history has had a greater impact on birth than bipedalism. Bipedalism has been linked with many other human adaptations, including expanding brain size and intelligence, tool using/making, hunting, highly dependent infants, dietary changes toward increased animal consumption, and increased energy efficiency. Whatever the "prime mover," the anatomical changes that accompanied the evolution of this unusual form of locomotion were significant and almost every one had an impact on birth because they all influenced the shape of the passageway through which almost every baby is born.

A full description of the anatomical changes that were associated with the transition to bipedalism is probably not necessary, but there are a number of important adaptations that are revealed when the pelvises of a two-legged (biped) and four-legged (quadruped) animal are compared (see Figure 5-1). These include, for the biped, an overall narrowing of the birth canal and pronounced spines that protrude into the birth canal.[1] The hominin sacrum is also broader, which serves to provide stronger and better support for the upright trunk. The top of the sacrum (sacral promontory) also protrudes forward into the birth canal,[2] reducing the opening size further and constricting the back portion. The actual birth canal or passageway has modified itself from a quadrupedal "shallow bony ring" to a bipedal "deep curved tube."[3] These changes are related to muscle alterations for successful

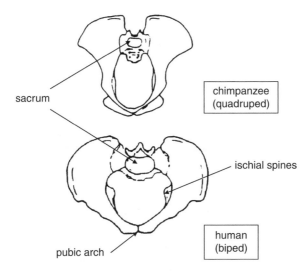

Figure 5-1 Comparison of a chimpanzee quadrupedal pelvis with a human bipedal pelvis, emphasizing features related to birth.

striding and maintaining balance above the center of gravity associated with bipedalism.

Human males and females are physically different in a number of ways, a phenomenon known as sexual dimorphism. This is most clearly seen in overall body size, secondary sexual characteristics (breasts, buttocks, facial hair), distribution of body fat, and degree of muscularity. For osteologists, however, the most reliable part of the skeleton to use for assessment of sex is the pelvis (see human pelvis in Figure 5-1). This is because the pelvis, in addition to its functional role in locomotion, also reflects modifications for childbirth, which are, of course, restricted to females. In general, efficient bipedalism is best with a narrow pelvis, but successful birth is best with a broad pelvis. The female pelvis that has evolved in response to these competing demands is, in some sense, compromised on both accounts: women are not quite as accomplished at bipedal running as men are (the fastest runner in the world will forever be a man), nor is delivery of an infant as easy for her as it is for our close relatives the chimpanzees. In an evolutionary sense, she has traded high-speed bipedal running for the ability to deliver large-brained infants.

At first glance, it seems that the slight sacrifice in efficiency that a wider pelvis would afford would not be as costly as the risks that the narrow pelvis places on childbirth. Why not just grow a very wide pelvis that makes childbirth easy? One can invoke the need for speed in escaping predators, but even that does not seem like a powerful enough selective force. While it may be true that a healthy woman in reasonably good physical shape is a pretty good biped, when we consider that adult females in the evolutionary past spent most of their adult years either pregnant or carrying infants, the selective advantage of efficient bipedalism becomes more apparent.[4] In the latter stages of pregnancy and early months of nursing and carrying infants, energetic demands on females increase significantly, placing greater selective emphasis on efficient bipedalism. The resulting balance is a pelvis that allows for a high degree of energetic efficiency in locomotion but also allows a large-brained infant to pass at birth, albeit with a tight squeeze.

Here is what an osteologist looks for in determining if a pelvis is female (see human pelvis in Figure 5-1): greater angle and concave curvature in the area under the pubic arch; wider birth canal in both side-to-side and front-to-back dimensions; greater sideways flare to the hipbones; a more rounded pelvic opening when seen from above (the typical male pelvic opening is heart-shaped); ischial spines on the side rather than toward the front; a top of the sacrum that does not project into the birth canal; and a slightly broader sacrum with less curvature. Certainly these characteristics are highly variable from woman to woman and depend on her overall body size, activity levels, and health and diet during growth, but if several are present, they indicate femaleness, and, of course, they all serve to enlarge the birth canal and reflect adaptation to childbirth. As obstetrician Maurice Abitbol[5] has pointed out, the characteristics of the pelvis that distinguish humans from chimps are those that are attributable to bipedalism. Those that distinguish males from females are attributable to increased brain size (encephalization) and the requirements of childbirth. From the perspective of human evolutionary history, the earliest hominin pelvis (australopithecine) was adapted to bipedalism and the more recent (*Homo* species) is adapted to birth of a large-brained infant.[6]

Our discussion so far is to distinguish quadrupedal from bipedal pelvises and male from female pelvises, and it neglects the great variability seen in pelvises among women. Most medical research has been done on healthy women who grew up in reasonably good environments with adequate nutrition and health care.[7] Their pelvises tend to be well developed for childbirth. For women who grow up in poverty with poor nutrition and overall poor health, however, pelvic dimensions described in medical textbooks and "birthing" books may not describe them. Also, taller women tend to have pelvises that better fit the babies they bear, whereas cephalopelvic disproportion or CPD (when the head is too large for the pelvis) is more common in shorter women.[8] There may be a genetic component to short stature, but as we have seen, it is more often due to undernutrition and poor health during childhood. (It should be noted that there is equivalent variation in male pelvises, whose shapes are also affected by nutrition and health factors.)

Birth in Our Ancestors

Fossilized skeletal remains of early hominins have been extensively studied to ascertain how and when bipedalism evolved. Although there are very few pelvic remains from the earliest group, known as australopithecines, there are two almost complete pelvises and, as luck would have it for our interest in birth, they both appear to be from females.

The major changes in the pelvis occurred when the lines leading to modern chimpanzees and humans diverged, approximately 5–7 million years ago. Birth is not "easy" for most primates, given the close correspondence between the size of the neonatal head and the maternal bony pelvis[9] (Figure 5-2), but bipedalism placed even more constraints on the birth process through alteration of the dimensions of the pelvic entrance and exit. For quadrupedal primates, the birth canal has its greatest breadth in the front-to-back dimension, but with bipedalism, the birth canal is "twisted" in the middle so that the greatest dimension of the entrance is side-to-side and the greatest dimension of the exit is front-to-back. Because the passage is such a tight fit, the neonatal head must line up with each of the dimensions, meaning that the baby enters the birth canal with its head facing to the side and rotates approximately 90 degrees in the middle so that the head lines up with the broader front-to-back dimension of the exit.[10] One way to envision this is to make a fist with your hand. The side-to-side (thumb to little finger) dimension is most likely greater than the front-to-back dimension. Now imagine a narrow oval passageway that is twisted in the middle so that the entrance is perpendicular to the exit. If you envision pushing your hand through the passageway, you will see that it has to rotate to pass through. This is pretty much what a baby's head has to do to get through the birth canal.

But it is not just the baby's head that has to pass through the birth canal. Following the head in a vertex (head first) presentation are the shoulders.[11] Unlike

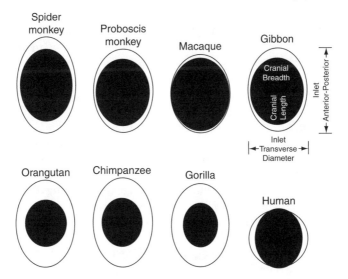

Figure 5-2 Pelvic openings (outer circles) and neonatal heads (dark inner circles) in selected primate species. From Schultz, 1949 and Jolly, 1985.

monkeys whose shoulders are somewhat narrow (a good adaptation to running on all fours) and can collapse when passing through the birth canal, humans have broad, somewhat rigid shoulders that are not as flexible. They are also perpendicular to the long dimension of the head; so not only does the head rotate in the middle to line up with the long axis of the pelvic outlet, it must rotate again as it exits the birth canal to enable the shoulders to pass through the transverse dimension of the inlet. Once the head has emerged, it undergoes further rotation to bring the shoulders through the pelvic outlet. With large babies, a phenomenon known as *shoulder dystocia* in which the shoulders get stuck at the pubic bones may occur following emergence of the head. This can be life threatening, although it may not have been a significant cause of mortality in the past because infants were unlikely to have been large at birth. I will discuss this in more detail in the section on birth complications.[12]

Additionally, adaptation to bipedalism has resulted in the top of the sacrum (sacral promontory) protruding into the birth canal so that the front of the maternal pelvis is more spacious than the back. Because the back of the baby's head is the broadest, those two dimensions line up so that the baby usually emerges from the birth canal facing toward the mother's back. If your image of childbirth is of a woman lying on her back on a bed or delivery table, this means that the baby is born facing the mattress or pad rather than facing the ceiling (midwives I have worked with call this "sunny side up"). In quadrupedal primates, the back of the baby's head lines up with the back of the mother's birth canal and passes straight through without rotating to emerge facing the front of the mother's body. When the baby's head emerges from the birth canal, the mother monkey or ape reaches down and guides it out along the normal contours of its body (Figure 5-3).

In humans, however, the mother often finds it difficult to reach behind her or between her legs to complete the delivery and risks pulling the infant against the angle of flexion in a way that damages nerves and muscles (see Figure 5-4). I have argued that the tendency for the human neonate to be born facing the mother's back accounts for the almost universal practice of seeking assistance at birth.[13] For most primates and most mammals, birth is a solitary event and females typically seek isolation rather than companionship when they begin labor. Human females, in contrast, tend to seek companionship at birth in almost all cultures. Certainly it is possible for women to give birth unattended, but my anthropology colleague Karen Rosenberg and I have argued that throughout human evolutionary history, those who sought assistance at birth had more surviving offspring than those who delivered their infants alone.[14] Even a small reduction in mortality and morbidity rates over hundreds of generations could account for the near-universal practice of accompanied birth that we see today.

Our ancestors probably sought out others not because of a conscious awareness that it would reduce mortality; instead, it is more likely that they felt anxiety and uncertainty about labor and delivery and sought companionship for emotional support. For most of human history, those emotional needs at birth have been

Figure 5-3 Delivery of a monkey infant and the mother's response. Redrawn from Trevathan, 1987.

met by friends and family members, most of whom were probably women. These people may not have had midwifery skills, but they cared about the woman and her newborn and were able, simply by being present, to reduce emotional stresses that could interfere with labor and delivery. But today, many women give birth in unfamiliar surroundings (in hospitals or birth centers), often with people they do not know, and their emotional needs are often not met. Accompanying that experience are fear and anxiety in labor that are considered to be problems (or "defects" in the vocabulary of evolutionary medicine) that need to be dealt with medically (with pain-relieving drugs, for example), rather than the "defenses" that they once were when they led women to seek the companionship and assistance at delivery that reduced morbidity and mortality. Of course, when anxiety and fear become extreme, medical intervention may be necessary for a healthy birth

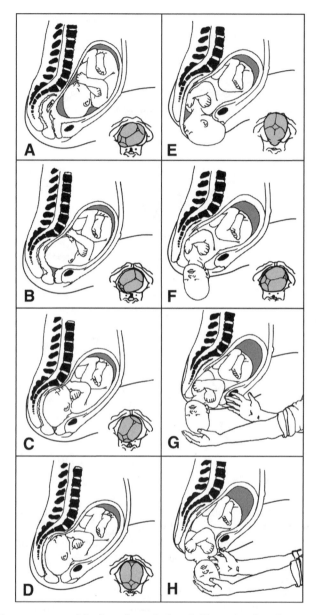

Figure 5-4 The emergence of the fetus from the bipedal human pelvis showing the benefits of having assistance at delivery. From Rosenberg and Trevathan, 1996.

outcome, but companionship may prevent the anxiety from becoming extreme in the first place. An evolutionary perspective argues that the anxiety many women feel at birth is a legacy from our ancestral history of a time when it helped to reduce mortality. In Randy Nesse's words, "what good is feeling bad" about birth is increased reproductive success[15]; in other words, women who felt "bad" (anxiety) about giving birth alone may have had more surviving offspring.

Brain Size at Birth

Another consequence of altered pelvic shape in the evolution of bipedalism is that an upper limit was placed on the size of the neonatal head that could pass through at birth. This meant that the birth process was in direct opposition to selection for increased brain size in human evolution and some sort of evolutionary compromise was required. (Obstetrician Philip Steer calls this a "conflict between walking and thinking."[16]) Expansion of the birth canal to allow delivery of a large-brained infant would have meant that women would have sacrificed efficiency in bipedalism. As noted earlier, while this may not be a problem today with low activity levels and motorized transit systems, females who could not walk great distances carrying food and infants because of inefficient bipedalism would have been at a selective disadvantage. The only way for the conflicting trends of a narrowing of the pelvic opening and increasing size of the brain to be resolved was for more and more brain development to be postponed until after birth.

For most mammalian species, approximately half of brain growth has occurred by the time of birth. In most primate species, however, more growth takes place after birth than before and infants are born with less than half of their expected adult brain size. Although there are virtually no fossilized neonatal hominins, anthropologists Jeremy De Silva and Julie Lesnik[17] have developed a method for estimating neonatal brain size based on cranial capacity of fossilized adult skulls and statistical analyses of neonatal and adult brain sizes of contemporary primates. Their conclusions indicate that, among other things, modern humans have "precisely the brain size at birth expected" for a primate of our body size. What is notable about humans is not how large the brain is at birth but how much more the brain grows *after* birth. The percentage of brain growth that is completed at the time of birth gradually decreases from the earliest australopithecines to modern humans, related most likely to the increasing size of adult brains over that same period.

The great amount of brain growth that occurs after birth means different things for the infant and the parents. For the mother, energetic investment is greater postnatally than prenatally (pregnancy "costs" an extra 300 calories per day, while lactation "costs" an extra 500 per day beyond her normal needs). And the only way that more dependent infants could survive would be for mothers and fathers to invest more heavily in them in the first several months after birth while the neurological systems necessary for independent function develop.

Delayed development of the brain enables birth to occur through the narrow bipedal pelvis. Another characteristic related to undeveloped fetal brains that makes birth easier is that the cranial plates are not fused and can slide over each other at the time of birth in vertex presentations. Some of the ligaments that hold the mother's pelvis together can also relax a bit at the time of delivery under the influence of the aptly named hormone relaxin. Quadrupedal monkeys experience a lot of relaxation of the ligaments and joints of the pelvis at the time of delivery, easing the passage of the large-headed infant through the somewhat small birth canal. For bipedal humans, however, too much pelvic relaxation provides risks for the muscles and structures that maintain bipedal locomotion, so the degree of flexibility at delivery is limited.[18]

Another unusual characteristic of modern humans that may contribute to successful childbirth is that the place where the two pubic bones come together does not fully fuse at the end of adolescence as it does in other primates. This means that the pelvic opening can continue to expand even after growth has ceased in the rest of the skeleton. In fact, the pubic bones do not fuse until after the most fecund years (until the early 30s)[19] are over. This characteristic has not been observed in australopithecines, suggesting that its evolution is related to birth and increased brain size rather than bipedalism.

Most of the discussion so far has revolved around selection for less brain maturity at birth so that the baby's head can get through the tight squeeze of the bipedal birth canal. Certainly this crude mechanical constraint played an important role, but it appears that there are numerous advantages for the infant that derive from being born before growth is completed. One of the most significant has to do with language learning at a time when the brain is still developing.[20] In a general sense, being developmentally immature at birth provides greater flexibility in cognitive development overall.[21] Infant motor immaturity may seem like a disadvantage at first because the energetic costs to the mother of carrying the young are significant, but it sets the stage for the development of complex cognitive and social skills. Arguably, the world outside the womb is much more stimulating for infant learning than inside the womb, although that is not to say that prenatal learning is not important or extensive.

Medical Consequences of Bipedalism

As discussed in the introductory chapter, the medical consequences of bipedalism range from minor back problems to pelvic organ prolapse. But nothing can match the impact that bipedalism has had on childbirth, an impact that was probably a major source of mortality before about 500 years ago and therefore a major agent of natural selection. The professions of obstetrics and midwifery probably would not exist were it not for the inherent conflict between the bipedal pelvis and the large-brained infant.

Pelvic Organ Prolapse (POP)

In addition to serving as the passageway for birth, the pelvic basin also serves as a "bowl" that supports the internal organs in upright posture. In quadrupedal mammals, the tough muscles of the abdominal wall support the internal organs, slung beneath the spine as the animals move or stand. In humans, the ischial spines, which are often the most challenging aspects of the pelvis for safe delivery, function to support the pelvic floor in a horizontal orientation and they protrude into the birth canal. Unfortunately, this horizontal position of the pelvic floor also makes it vulnerable to pelvic organ prolapse (POP).[22] The ischial spines provide another example of an evolutionary trade-off: they support the muscles and ligaments that hold the pelvic organs, but they protrude into the birth canal at its most narrow point, often interfering with delivery.[23] As another way of looking at it, the ischial spines and muscles of the perineal area have been "adapted for holding things in rather than for letting them out." [24]

Pelvic organ prolapse is a much more common problem for women than for men, due in part to the expansion of the pelvic opening for birth; it is especially common in older women who have given birth several times. It occurs when the pelvic floor weakens or is damaged and can no longer support the internal organs. The most likely culprit is damage due to overstretching of the levator ani muscle during birth; this is the muscle that wagged the tail in pre-hominin ancestors and now forms part of the pelvic floor.[25] Heavy lifting and chronic coughing or straining can also contribute to weakening. Pelvic floor prolapse is not usually life threatening but can cause pain and discomfort and may require surgery for relief. It also interferes with sexual functioning. A recent survey determined that in the United States, almost a quarter of women suffered some form of pelvic floor disorder, most prominent of which are incontinence and organ prolapse.[26] As another example of trade-offs in evolution, women with wider pelvic inlets and shorter birth canals, which make for easier birth, are at greater risk.[27] Women whose pelvic inlets are very wide (called a platypelloid pelvis in the medical literature) are at greater risk for pelvic floor disorders than those with narrow inlets ("anthropoid pelvises".)[28]

Bipedalism requires the pelvic floor to support the pelvic organs, including the bladder, rectum, and so on, but it also requires changes in processes related to elimination. A narrow bony pelvic ring helps with the support function. Rather than being formed primarily by muscles, the bulk of the pelvic floor is made up of strong connective tissue, which provides better support in that it does not require muscles to constantly constrict the sphincters of the anus and bladder to keep them closed. Bipedalism requires a "more refined control of the visceral apertures."[29] In summary, the pelvic floor has evolved from a tail-moving function in quadrupeds to a weight-supporting and sphincter-controlling function in bipedal humans.[30] It is probably not a stretch to say that constipation, incontinence, hemorrhoids, and herniation are all by-products of bipedalism.[31]

Furthermore, the relatively small pelvic basin also holds the uterus and fetus during pregnancy. In early pregnancy, when the fetus is still small, this is not much of a problem, but in later pregnancy, as the fetus grows, it "spills out" of the basin and protrudes into the abdomen. The abdominal wall is fairly rigid and does not allow much extension. There is a limit to how large the fetus can grow, a fact that may partially explain why multiple gestations tend to be shortened—the combined sizes of the twins or triplets exceed the capacity for the abdomen to hold them toward the end of pregnancy. Expanding fetal size in pregnancy also puts pressure on the blood vessels, which compromises both maternal and fetal health. Significantly, the forward protrusion of the fetus into the abdomen also changes the center of gravity for the woman's body and provides challenges to maintaining upright posture while standing and walking, as noted in the previous chapter.[32] For pregnant quadrupeds, where the fetus is balanced among the four legs, this is not a problem. So it is not just birth that bipedalism challenges, but the latter part of pregnancy also.[33]

Cesarean Section

Another medical consequence of bipedalism for some women is the need to deliver an infant via cesarean section. As noted, birth is another point when the interests of the mother and the fetus may not be the same. Larger babies (up to a point) are more likely to survive the birth process and the neonatal period than small babies, but their delivery is much more difficult for both the mother and the infant. It is in her best interest to give birth to a sufficiently large infant that can survive, but not so large an infant that the mother's own health is compromised. The optimal birth weight for the infant (the weight at which perinatal mortality is lowest) is between 3,800 and 4,200 grams, according to studies in health-rich nations.[34] There are no data reporting ease of delivery for birth weight categories, but it is probably safe to say that the smaller the baby, the easier it is to deliver. Of course, the mother's reproductive success depends on giving birth to a healthy baby, so it is clearly not in her interest to give birth to an infant too small to survive, no matter how easy it is for her to deliver. In fact, the average birth weight is less than the birth weight that is optimal for infant survival,[35] suggesting that the mother's needs override those of the infant. This makes sense in that in the past, if the mother died, there was very little chance that the infant would survive, but if the infant was a bit smaller than the optimum, his or her survival chances were still pretty good. In other words, the costs to the infants are less than the costs to the mother in this case.

Babies that are too large are often delivered by cesarean section, which may increase the infant's survival chances, but the mother's risk is greater with surgical delivery than with vaginal delivery. Of course, in the past, both mother and infant would have probably died in the absence of surgical delivery options. Infant birth weight is an excellent example of stabilizing selection—higher mortality is associated with weights too large or too small, so natural selection keeps birth weight balanced between about 2,500 and 5,000 grams. To some extent, however, the constraints against birth weight have been removed by technology that keeps alive very low birth weight babies and allows cesarean section for very large babies. It would not be surprising to see a gradual increase in the mean birth weight in some populations now that overall health and nutrition have improved and surgical deliveries are possible. In fact, in the United States over an 18-year period, average birth weight increased by 40 grams, a 1.2% increase. National U.S. data from 1960 to 1997 show a 2% increase in babies weighing 3,500–3,999 grams and a 1% increase in those weighing 4,000–4,999 grams.[36] This indicates that the average birth weight is drawing closer to the optimum because the risks to the mother of having a large baby have been reduced by c-section. There is even the suggestion that if this continues, virtually all babies will have to be born by c-section in the future.[37] For example, it appears that the English bulldog, bred for squat bodies and large heads, has reached the point that virtually all are delivered by c-section.

Cesarean section rates are on the increase worldwide (29% in the United States; 23% in the UK; 22% in Canada; 32%–70% in Brazil; 23% in Australia; 35% in Italy and Taiwan; 40% in Chile). Most investigators suggest an "expected" rate of 10%–15% for c-section and this is the rate recommended by the World Health Organization. Risk factors for surgical delivery include short stature, obesity, contracted pelvis (usually from poor nutrition during growth), and infection. It is possible that the "natural rate" may be increasing because of undernutrition during growth that results in short stature and, surprisingly, obesity in adulthood. For example, in Western Australia, women shorter than 160 cm had a c-section rate four times that of women taller than 164 cm.[38] Women with a body-mass index (BMI) above 30 have a three times greater risk of c-section than women with BMI below 20.[39] Obese women tend to have obese babies, which may explain the higher c-section rates. A final "natural" reason for increased c-section rates may lie in maternal age, which tends to be later in most of the countries with high rates.[40]

Of course, risks of cesarean section are still present and include death (seven times the risk of vaginal delivery in a Netherlands study), hemorrhage, pulmonary embolism, sepsis, and anesthesia complications.[41] Recovery time is longer and breastfeeding and mother-infant bonding are often compromised. Cesarean section may also compromise a woman's future fertility. For example, there is evidence that stillbirths increase in pregnancies following a c-section, as do the rates of placental problems.[42] It is also common for women who have one cesarean section delivery to have all future children this way, whether indicated or not. Furthermore, if a woman has a c-section for cephalo-pelvic disproportion (CPD) or dysfunctional labor, the chances are higher that her daughters will also require c-sections.[43] Maternal-fetal conflict theory might argue that a big baby also reduces competition for resources from future siblings when the mother's chances of having more children are reduced.

Cesarean section also presents problems for infants, including higher incidence of respiratory distress, disrupted sleep rhythm in the early postnatal days,[44] and greater challenges to self-regulation, the process by which the infant begins to maintain behavioral and physiological balance following the disruptions caused by birth. Because the mother has experienced major surgery, she may not be in a position to help her infant achieve self-regulation through the usual interactive processes of holding, stroking, skin-to-skin contact, and vocal and visual engagement. In one study of women two to four weeks after delivery, the brains of mothers who delivered vaginally were significantly more responsive to their babies' cries than those who delivered by c-section as shown in fMRI (functional magnetic resonance imaging).[45]

Some of these complications may be due to drugs that the mother receives during surgery, but there also may be disadvantages for the infant who "skips" labor. Although we are just beginning to ask these questions, there appear to be long-term effects of cesarean delivery on the infant that may cast doubt on the advisability of elective c-sections. A particularly alarming review concludes that the risk of developing type 1 diabetes is 20% greater in children delivered by c-section and that no other known factors can account for this increase.[46]

On the other hand, cesarean section has saved lives of countless mothers and infants in cases of cephalo-pelvic disproportion (CPD), malpresentation, placenta previa, detached placenta, cord prolapse, uterine rupture, fetal heart rate abnormalities, and severe infections.[47] Cesarean sections are increasingly recommended for women who are HIV positive or who have active outbreaks of herpes simplex virus (HSV); benefits of c-section for other infections like hepatitis B virus, human papilloma virus, and hepatitis C virus are more controversial.[48] Urinary incontinence is less problematic for women who had c-sections. For women who have extreme fear of childbirth, the option for surgical delivery is comforting. Finally, it is often more convenient to have a scheduled, elective cesarean delivery for women who find their lives are too complicated for "natural" childbirth. Hospital births with electronic fetal monitoring, induction and augmentation of labor, epidurals, forceps, episiotomies, and unfamiliar attendants hardly seem more natural or desirable to women who have the option of a scheduled surgical delivery.[49] Furthermore, the type of delivery with the greatest risk is an emergency cesarean section, which a planned c-section can avoid, especially if the baby presents breech or is past term or if the mother had a previous c-section.

Is Birth "Good for" Babies?

As noted, an increasingly common technological way of dealing with the tight squeeze of the baby through the birth canal and the anxiety and fear that some women feel is to deliver the baby via cesarean section. But with the rapidly rising rates of cesarean section and concerns about maternal and infant health following what is undeniably major surgery, one might wonder if there are advantages to being born vaginally. In other words, is there a benefit to being pressed by uterine contractions, navigating the tight quarters of the birth canal, occasionally being deprived of oxygen, and having your head compressed? Anthropologist Ashley Montagu was one of the first to suggest that relatively long labors were "good for" human infants.[50] In fact, he argued that the stroking and massaging provided by the uterine contractions replace the functional significance of licking the young seen in virtually all other mammalian species.

Among mammals, licking newborn infants stimulates development of the respiratory and digestive systems and infants that are not licked often die because of complications in these systems. The stimulation that the human baby's skin receives during labor also influences organ development and proper functioning of the nervous system. Montagu notes that respiratory and digestive problems and poor control of the bladder and sphincter are common ailments in infants born prematurely or by cesarean section, those who experience brief or no labor contractions.[51] There is evidence that infants born by cesarean after a trial of labor have better functioning systems than those born without the benefit of contractions.

Another reason that labor might be good for babies is that the stress hormones that both the mother and fetus produce during labor trigger the production of catecholamines (adrenalin/epinephrine and dopamine) in the infant that help it cope with life outside the womb. Specifically, these "stress hormones" increase production of agents that speed up maturation of the lungs and enable the lungs to expand to keep amniotic and other fluids from filling them up. They also serve to increase fetal blood flow, especially to the brain; increase availability of calories to the baby; and increase white blood cells for immune protection.[52] Further, catecholamines help the fetus withstand hypoxia during delivery.[53] They appear to be important for promoting breathing immediately after birth, and infants born by elective cesarean section without labor (and thus not producing the high levels of catecholamines) often have breathing problems. Blood sugar levels are low in babies delivered by elective cesarean section, reinforcing the significance of catecholamines for mobilizing energy.[54] This may be particularly important for infant survival in the first few days after birth in cultures that believe mothers should withhold the first secretions from the breasts (known as colostrum) and delay breastfeeding until the milk comes in.

Some have argued that the stress hormones also play a positive role in mother-infant bonding.[55] Infants who are born after unmedicated births are often very alert for the first few hours after birth, perhaps due to catecholamines such as norepinephrine, which surge in labor. This alertness and ability to respond to stimuli play a role in initiation of mother-infant interaction. Infants delivered by cesarean section, however, do not show this level of alertness nor do they have high levels of norepinephrine in their blood.[56]

Norepinephrine facilitates development of the neonatal olfactory system, so one of the pathways through which this might work is olfaction: babies who were born after normal labor and delivery showed enhanced learning of odors to which they were exposed immediately after birth compared with infants born by cesarean section.[57] Babies who experienced labor contractions before surgical delivery showed better olfactory learning than those who were delivered before labor began.[58] This suggests that infants can more readily recognize their mothers' odors if they have been through the contractions of labor, which may contribute to maturation of the olfactory system. The catecholamine surge during delivery also dilates the pupils and increases alertness, both of which contribute to bonding.[59]

The birth process also appears to initiate an acute phase response in the infant, which serves as the infant's first line of defense against infectious agents encountered in the new environment.[60] Part of the response is an increase in body temperature, which is extremely important for neonates who are unable to effectively regulate their body temperatures. In the ancestral past, this ability to increase temperature may have enhanced survival. Infants delivered by cesarean section produce much lower levels of the immune agents that are part of the acute phase response. This and other aspects of the early postpartum period will be discussed more extensively in the next chapter.

The view of elective cesarean sections from evolutionary medicine argues that it may be possible to have the benefits of labor and the benefits of cesarean section when necessary. Babies who are delivered surgically after labor has been initiated show a reduced risk of respiratory problems.[61] Similarly, as noted earlier, babies who have experienced labor show higher levels of catecholamines even if delivered by c-section compared with those who did not experience labor. So allowing labor to begin before performing a cesarean section may enable the mother and baby to have the advantages of both.

Obstructed Labor

When the fetal head is too large to pass through the birth canal or is in an unfavorable position, a medical condition known as obstructed labor or dystocia results, leading to an emergency c-section in hospital deliveries. Approximately 8% of maternal deaths worldwide are due to obstructed labor; for the infant, stillbirth, asphyxia, and brain damage may result. If the mother survives, a long-term consequence may be an obstetric fistula, which is an opening between the vagina and the bladder or the rectum. Unless it is surgically repaired, the consequences to the woman include an inability to control urination and defecation, chronic infections, pain, infertility, and even death. It is estimated that more than 2 million women in the world are currently affected by obstetric fistulas and many are abandoned by their husbands, families, and communities. Embarrassment often keeps the affected women from seeking help, even if it is available.

It seems untenable, in an evolutionary sense, for obstructed labor to have been a major cause of death in past populations and it appears to be rare in wild animal populations. Perhaps obstructed labor is a phenomenon that may be related to modern affluence, more-than-adequate nutrition, and reduced activity levels. Robert Roy,[62] an obstetrician who practiced among the Canadian Inuit, notes that not only was obstructed labor extremely rare among these women, but labor and delivery appeared to be rapid and easy. The c-section rate reported in 1978 was less than 2% (10 of 622 Inuit deliveries, only 4 of which were for CPD). Similar reports come from observers of other foraging populations, but as I have noted previously,[63] reports of easy deliveries are usually based on interviews with people other than women who have given birth and they also may come from cultures in which the cultural ideal is to be stoic during labor and delivery, so the concept of "easy delivery," while perhaps true, must be interpreted with caution. That having been said, it seems likely that the relatively poor diets often associated with agriculture may lead to smaller bodies with small pelvises and greater difficulty giving birth, so it may be true that obstructed labor was relatively rare until the origins of agriculture, 10,000 years ago.

Another reason we may see more evidence of obstructed labor in health-rich populations has to do with the posture that a woman typically assumes during delivery in most hospital births: flat on her back in a position known as *lithotomy*. It is not hard to see that this is probably the least optimal position for birth in comparison with squatting, sitting upright, or even standing. For one, gravity is working against the fetus when the mother is lying flat on her back. Perhaps this position is best for birth attendants, but it does not appear to be in the best interest of mothers and babies. The most common position for delivery reported in the anthropological literature is seated or semi-reclining, although squatting, kneeling, and standing are also frequent. The flat-on-the-back position is rarely used in other cultures or in nonhospital births in health-rich nations.

In an upright position, the force of the uterine contractions is born by the back of the baby's head (the occiput) in a normal head-first position.[64] The occiput is the most developed of the fetal cranial bones and is the best for withstanding the intense pressure of the contractions. When the mother is lying down, it is the baby's fragile frontal bones that have to bear the force against her sacrum. Of even more concern is that if the infant is lying with the back of its head (the occiput) against the mother's sacrum (known as a posterior presentation) the force of the contractions is applied at the base of the skull near the spinal cord.

A recent study of the effect of position on pelvic flexibility has confirmed that the birth canal widens in several dimensions in the standing or squatting positions in comparison with the flat-on-the-back position.[65] Unfortunately, for women who hope to deliver in the squatting position, it is often extremely difficult if they have not spent a lot of time in that position in their everyday lives. In cultures where women routinely squat in delivery, they also squat around cooking fires, for elimination, and in social interactions, so they have the muscle development to support their weight during delivery. If someone is available to support her, however, even a woman who does not routinely squat can deliver in that, potentially more optimal, position.[66]

As noted earlier, a relevant dimension of the fetus that is often overlooked is the shoulder breadth. The incidence of shoulder dystocia, whereby the shoulders get "stuck" in the birth canal, is approximately 2% in infants weighing between 2,500 and 4,000 grams but rises to almost 10% for infants greater than 4,000 grams.[67] It occurs more commonly in women with diabetes, which suggests that its incidence may rise in the future as rates of diabetes rise worldwide. By-products of shoulder dystocia include hemorrhage, major perineal lacerations, uterine rupture, cord compression (and thus fetal hypoxia), and even fetal death. If efforts to maneuver the shoulders from the stuck position do not work, attendants may resort to breaking the collarbones so that the shoulders can "collapse." The mother's posture during birth is one important variable in preventing shoulder dystocia. Squatting or standing, because it helps to widen the pelvic outlet, makes shoulder dystocia much less likely than the flat-on-the-back position.[68]

There are other aspects of the way in which the medical community treats childbirth that may have negative impacts on women during labor and delivery and contribute to obstructed labor.[69] Obstetrician Michel Odent is probably the foremost spokesperson for the argument that difficult labor and delivery are almost exclusively products of modern medical treatment of birth.[70] In his view, difficult birth is a "disorder of civilization," which results from the extreme anxiety that is induced by the medicalization of birth. He argues, in fact, that birth needs to be "dehumanized" by allowing it to proceed as it does in other mammals, without any technical or social intervention. To him, the ideal birth takes place alone with no intervention whatsoever and with the woman following her own instincts. In these cases, he proposes that a "fetal ejection reflex" will allow delivery to proceed in the way that it does in other mammals who seek isolation when labor contractions begin. Even considering Odent's arguments, I still maintain that in most cases today birthing women benefit from social and emotional support, but his methods may work well for some highly motivated women.

Birth Does Not End with the Delivery of the Placenta

One of the most important lessons of evolutionary medicine is that birth is not a discrete event but is part of a process that begins in the developing ovary of a several-week-old fetus and ends only with independence and the onset of the next cycle of reproduction. In fact, from an evolutionary perspective, it is odd that in modern medicine, one professional (a midwife or obstetrician) takes care of the woman and her fetus during pregnancy, labor, and delivery and another professional (neonatologist or pediatrician) takes care of the baby—and even odder that there does not seem to be a parallel profession that takes care of the mother after birth.

Anthropologists beginning with Ashley Montagu[71] and Adolf Portmann[72] have argued that human infants are born at an earlier stage of development such that the first several months of life are more like gestation than independent living. Montagu used the concepts of *uterogestation* and *exterogestation* to distinguish the nine months in utero from the six to nine months outside of the uterus in which gestation continues, although in a different way. "We are therefore justified in thinking of the first months of a child's life as a continuation of the period of gestation, and we are justified in treating the child and his mother, and the relationship between the two, as though they were still closely connected."[73]

Another good reason to think of birth as a process that does not stop with the delivery of the placenta can be found in the data on maternal mortality related to childbirth. In 2005, an estimated 529,000 maternal deaths due to childbirth occurred in the world, a ratio of 400 deaths per 100,000 live births. About half of the deaths occur in the first 24 hours after birth and more than two thirds occur in the first week. Of course these deaths are not distributed evenly throughout the world and they are a relative rarity in health-rich nations such as the United States, Canada, Australia, New Zealand, and those of Western Europe.

The most common cause of maternal mortality related to birth is postpartum bleeding. When the placenta detaches from the uterine wall, it leaves behind a "wound" of open blood vessels that will continue to bleed if they are not closed. One of the functions of postpartum uterine contractions is to help the uterus "shut down" and close the open blood vessels. This usually proceeds without complications, unless there is a portion of the placenta or other membranes retained in the uterus that get in the way of closure. In hospital deliveries today injections of oxytocin are routinely given to inhibit postpartum hemorrhage, but in the past, what did women and their companions do to prevent excessive blood loss? Just as the mother and her birth companions do a lot to enhance the health and survival of the infant immediately after birth, it appears that the infant has ways of enhancing the health and survival of its mother. Within a few minutes after birth, an infant will begin to lick, nuzzle, or even suckle the mother's breast. These forms of nipple contact stimulate the release of oxytocin into her bloodstream, which results in uterine contractions, expulsion of the placenta, and closure of the open wound left behind by the placenta. If the infant is not able to move toward the mother's breast, stimulation of the nipple by any method available accomplishes the same thing, but it is compelling to think that the infant can save its mother's life by preventing postpartum hemorrhage in a very low-tech way.

We have seen that birth in humans is very different from birth in other mammals, even our closest living relatives. Many of the differences are due to changes in the birth canal that occurred as natural selection favored bipedal locomotion in our ancestors 5–7 million years ago. With increased brain size about 2 million years ago, more and more of brain development was postponed until after birth so that infants could pass through the narrow birth canal before their heads got too large. This meant that newborn human infants were quite undeveloped at birth, requiring a number of modifications of parenting behaviors. We have also seen the benefits from having assistance at delivery because the baby emerges from the birth canal facing "backward."

Even though I argue against treating birth and the immediate postpartum period as separate events, I have done just what I advocate against by ending this chapter with birth and beginning a new one with infancy. To do it differently would make for a long chapter, so read on as if we are just taking a breath rather than opening a new chapter in the life cycle.

6

The Greasy, Helpless One-Hour-Old Human Newborn

Don't let anyone tell you that the first hour after birth is not important for mother-infant bonding. Don't let them tell you it is *necessary* for a mother and infant to be together during that time, either. How can both of these be true? The first statement pertains to our evolutionary legacy as mammals with evolved behaviors that assist in formation of a secure bond that enhances infant survival. The second statement is true because of our legacy as an incredibly flexible species whose dependence on evolved mechanisms for love and attachment is minimal. What is so special about the first hour or two after birth? To understand this, it is important to think about childbirth in the past, before high-tech equipment was available to help the baby adjust to life outside the womb and to facilitate recovery and stabilization of both mother and infant following birth. For much of human history, the mother was the key to infant survival from the moment of birth onward and her presence during the first hour after birth was particularly important.

Infant Helplessness at Birth

As noted in previous chapters, human infants are born with a degree of helplessness not seen in most members of the Primate order. Across species, infant development can be described as *precocial* or *altricial.* Precocial infants are those that are well developed at birth, with open eyes and motor skills that enable them to follow or cling to their mothers. Examples include most hoofed mammals and many primates. Altricial infants are helpless at birth, usually do not have open eyes, and must be left in nests or carried by their mothers. Dogs, cats, and a few primate species are altricial. The two degrees of infant development are at one end of a continuum and most species fall somewhere along that continuum rather than at one extreme or another.[1]

Human infants are like the precocial monkeys and apes in many ways, but they are "underdeveloped" in other ways. The most obvious, and one we emphasized in the previous chapter, is that the brain is only about a quarter of the way through its growth trajectory at the time of birth, so many of the neurological functions that are tied to the immature brain are equally immature. Humans are often referred to as "secondarily altricial,"[2] because of these mixed features of altriciality and precociality.

I think of the concept of secondary altriciality as a metaphor for describing the evolutionary history of the human infant. The earliest primates were probably altricial like all of the early mammals. The more recent ancestors that we shared with monkeys and apes probably gave birth to infants that were as developed as monkey and ape infants today, that is, somewhat precocial. We also have milk composition that is more typical of precocial than of altricial species. Based on this evidence, I have argued that our ancestors became more altricial with the evolution of bipedalism and enlarged brains beginning about 2 million years ago, so the helplessness we see today in our infants is "secondary" to having evolved through a more precocial stage in the past.[3]

There is further evidence that humans are less developed than would be predicted based on our closest relatives. The degree of bone development seen in macaques at the time of birth is not present in humans until they are several years old.[4] As noted in the previous chapter, cranial plates are not as well developed in human infants in comparison to ape and monkey infants. This is advantageous for human birth because the bones of the front part of the skull can slide across one another, making the diameter of the head smaller and easier to deliver. Apes have powerful jaw muscles to enable them to chew the coarse, fibrous foods that make up their diets. If their babies had cranial plates that were as undeveloped as those in humans, their powerful chewing muscles could pull their skulls apart when they started eating solid foods.

Other systems that seem underdeveloped in comparison with other primates include gastrointestinal, immune, and thermoregulatory systems. Newborn infants lack enzymes for digesting foods other than colostrum and milk and will not develop them until they are several months old. This is one of the reasons that providing foods like cereals to very young infants can cause digestive problems. The American Academy of Pediatrics recommends not feeding cereals to young infants until they are four to six months old. As we have seen, newborn infant immune systems are immature and will not be fully functioning until the baby is about a year old. Before that time, the infant is dependent on the immune properties acquired from the mother during pregnancy and from breast milk. As we will see, there are several things that happen during and soon after birth that can have a positive (or negative) impact on the infant's immune function and that reinforce the idea that birth is part of a continuum rather than a discrete event.

There are a number of ways in which a mother maintains her infant's health throughout the prenatal and postnatal period that strengthen the notion that the 15–21 months following conception are a continuum of fetal-like development. For example, during gestation, fetal body temperature is maintained by the mother's thermoregulatory system. After birth, the newborn infant has a limited ability to maintain a stable body temperature, but the mother can continue to keep him warm with her own body. Pediatrician Jan Winberg describes the "nest" that the mother creates with her breasts, chest, and arms and notes that this nest provides more warmth to the infant than an artificially warmed hospital cot.[5] Comparing skin-to-skin babies with cot babies, he and his colleagues found that the skin-contact babies were better able to retain adequate blood glucose levels, a very important benefit before milk production kicks in. Skin-contact babies also cry less, which further conserves energy and would have kept the vulnerable mother-neonate dyad from being detected by potential predators in the past. When the infant suckles, the temperature in the mother's breasts and chests increases,[6] adding to the warmth. Certainly the mother's ability to help the baby conserve heat and energy was critical to survival in the past, as it still is in many areas of the world today.

A lot of the hormonal and physical changes that occur in mothers and infants at the time of and immediately after birth promote not only healthful adaptation to changed circumstances but also the development of a strong mother-infant bond.[7] Humans are not really all that different from other primates and even rats when it comes to the tools available to enhance the ability of mothers to attach to their infants. These tools work to ensure that mothers will be willing to invest the huge amount of time and energy in caring for and raising young that will be dependent directly on them for sustenance and nurturance for several weeks (rats), months (monkeys), or years (chimpanzees and humans).

There has been wide-ranging debate about the importance of the first hour after birth (referred to as the maternal sensitive period by some scholars) for bonding, and I will not try to review the extensive literature here. Suffice it to say that birth and the associated changes in physiology, the sights, sounds, and smells, and the presence of the newborn infant all serve to focus the mother's attention on the infant and it is indefensible to claim that the hour after birth is just like any other hour in a person's life.

Certainly the hormones that lead to attachment in other animals probably contribute to bonding, but it is clear that they are not *necessary* for bonding to take place. Here is what I wrote about the proposed maternal sensitive period in 1987, and my thoughts about the first hour after birth have not changed in the interim:

> Consider the intense physical and emotional experience of giving birth and the hormonal actions that accompany the process. The mother has been aware of the existence of this child for several months. She has felt its movements and may have even talked to it. The newborn infant is finally present in living, breathing form, and, like a wrapped package that is explored upon opening, the mother is especially interested in finding out about it. . . . we all know the excitement of unwrapping gifts and the sense of discovery felt upon opening a gift box. We also know the disappointment felt when someone else opens a package intended for us. The value and meaning of the gift itself, in the long run, may not

have anything to do with whether or not we actually opened it, but there is added pleasure in doing so. So it is with the newborn infant. The later, long-term relations may have nothing to do with . . . whether or not the mother and infant were together for the fist hour getting-acquainted period. They missed out, however, on a special period of excitement that *in itself* is valuable and meaningful.[8]

What Do Mothers and Infants *Do* in the First Hour after Birth?

Following routine hospital deliveries in the middle of the 20th century, mothers and infants did not do much of anything together in the first hour after birth, so the question that heads this section would have seemed meaningless to those experiencing or witnessing births. Other animals, however, are known to exhibit behaviors after birth that are called "species-specific" because they are not only predictable for a given species, but are slightly different for each species. As noted in the previous chapter with regard to cesarean section, one near-universal behavior seen in mammals is the routine licking of the infant by the mother immediately after birth. Maternal licking of the infant serves a number of functions: removing some of the material that covered the fetus in utero so that it does not inhibit breathing and heat retention; stimulating breathing, digestion, and elimination; removing odors that might attract predators; orienting the infant toward the nipple; learning of infant odors; and facilitating bonding. Observers of maternal licking in mammals report that it seems compulsive and mothers ignore everything else in their environment until the task is completed. This makes sense from a survival stance because infants that are not licked often fail to survive.

One important exception to maternal licking of the neonate is the human species. I have been unable to find any report of licking in the anthropological literature and assume that humans are a rare exception to the near-universal mammalian pattern. (Ocean-dwelling mammals, which do not appear to lick their infants, are another exception.) But what human mothers do is use their hands extensively in interacting with their infants in the first hour after birth. In my observations of 66 births that occurred in a homelike setting, I found that mothers engaged in a predictable pattern of touching their infants, beginning with cradling or encompassing the infant very soon after birth and moving on to palmar massaging and finally to finger exploration of face, hands, and extremities.[9] I interpreted these behaviors as enhancing warmth, respiration, and gastrointestinal development, just as licking does in other mammals.

Touch is extremely important in human interaction so it is not surprising that skin-to-skin contact between mother and infant immediately after birth has far-reaching consequences, including enhancing successful breastfeeding, hastening attachment, reducing anxiety, increasing confidence in caretaking, and improving mood.[10] The likely promoter of all of these benefits is the hormone oxytocin, the production of which increases when the infant suckles and is in skin contact with the mother. We will see in the next chapter that successful breastfeeding is extremely important for infant survival, so evidence that skin-to-skin contact between mother and infant in the first few minutes after birth is related to significantly greater breastfeeding duration[11] adds credibility to the idea that early postpartum interaction is important to long-term health of both mother and baby.

The Evolutionary Significance of Vernix

A further benefit to the infant of rubbing and massaging is that it spreads the *vernix caseosa* into the skin, which protects it from drying out, among other functions. Vernix caseosa ("cheesy varnish") is a creamy white fatty substance on the skin of newborn babies that serves to protect the skin from dehydration in the amniotic fluid. Although it has antimicrobial properties,[12] it is usually wiped or washed off immediately after birth in hospital deliveries as part of the process of cleaning the baby and removing unsightly birth fluids. In home births, however, it is seen as an important substance for protecting the baby's skin and it is often rubbed in. Midwives with whom I have worked are eager to get a dab on their own hands because of its superb moisturizing properties.

Vernix is derived from fetal skin cells and is comprised mostly of water, with lipids and proteins at about 10% each.[13] If we assume that "nature's cold cream" is best for babies, it suggests that the high oil percentages found in creams and oils manufactured for babies may not be ideal moisturizers. Scientists at the Skin Sciences Institute of Cincinnati Children's Hospital Medical Center in Ohio are working on developing a moisturizer with a formula similar to vernix for use with premature infants whose skin is poorly formed, and for diaper rashes in infants and eczema and psoriasis in adults.[14] Furthermore, vernix is high in vitamin E and the skin pigment melanin, both of which have antioxidant properties and may help protect the infant's skin from pollutants and ultraviolet (UV) light.[15] It has an odor and, perhaps, pheromones that are attractive to parents and may contribute to bonding. Additionally, its smoothness may be a magnet for rubbing and massaging.

The baby's skin forms in utero, but the process requires that the top layer on which the new skin is developing be dry. Vernix serves as a barrier to water in the amniotic fluid that enables the new skin to develop. This happens between 25 and 26 weeks of gestation, so an infant who is born before that time will not have the top skin layers. This means that the premature baby will have problems with heat regulation, water loss, and microbe invasion. Medical researchers have even suggested that the skin may be the limiting factor for survival for ever-younger premature births. (Moreover, the time it takes to develop mature skin is part of the answer to the question of why pregnancy lasts nine months.) A cream with the same properties as vernix would greatly advance care of premature infants. When the infant is born, the previous function of vernix to keep the infant's skin from exposure to too much water switches to minimize evaporation and heat loss. The unique composition of vernix and its high water content allow it to serve both functions and to change quickly with birth and exposure to air.[16]

When a baby is born, its skin is exposed to air, bacteria, and sunlight; the skin is an important mechanism for coping with the cold, dry environment that the baby meets at birth. Newborn infants (at least those born in hospitals) are usually covered with various types of bacteria, including *Staphylococcus, Bacillus,* and *E. coli.*[17] As noted, neonates have low immunity (except that derived from their mothers) and the vernix seems to serve as a "complex innate defence barrier" as well as a promoter of colonization by useful microbes.[18] It may also protect the fetus from the tar-like "stools" that are sometimes excreted (known as *meconium*) and the neonate from infectious rash-inducing agents in feces.[19]

Vernix is more prolific in babies born by cesarean section, suggesting that the contractions of labor and the tight squeeze through the birth canal in vaginal deliveries results in some of the cream being rubbed into the skin. An early suggestion for the function of vernix was that is served as a "grease" to help the baby pass through the birth canal. The fact that c-section babies have a lot of vernix seemed to support this but the least amount of vernix is seen in babies that are larger and beyond term at birth, ones that presumably would benefit from having the most vernix to grease the passage of their larger bulk.

No other mammal is known to have vernix, which suggests that the substance may be related to the relative hairlessness of the human infant. Of course, humans *do* have hair and while in utero, the fetus is covered with a fine hair known as *lanugo,* most of which has disappeared by the time of full-term birth. Although the function of lanugo is debated (some even suggest it is an "evolutionary holdover" and has no current function), it may work with the vernix to improve its water repellant properties. The movement of the hairs may also contribute to the development of normal heart rate, which, in turn, enhances growth mechanisms. It is also associated with oxytocin release and it has been suggested that the pleasure experienced by the fetus from the oxytocin motivates fetal movement, which further stimulates growth mechanisms.[20] The process continues in the immediate postpartum period when the mother strokes her infant and rubs the vernix, no longer surrounded by lanugo, into the skin.

As noted, vernix protects the fetus from dehydration in utero, but it also serves other prenatal functions. Perhaps most important are the antimicrobial properties that protect the fetus from mild intrauterine infections. Interestingly, the mucus plug that closes the cervix until late in labor has many of the same antimicrobial agents that are found in vernix, suggesting that they serve similar purposes.[21] Because mild infections can initiate labor, the ability to reduce their effects is an important adaptation for ensuring that fetuses are sufficiently mature before labor begins.

Another function in utero is that some of the proteins in vernix that are swallowed with the amniotic fluid may help in the maturation of the gastrointestinal system. It appears that the maturing skin and maturing lung interact during late pregnancy to facilitate availability of the amino acid glutamine which is abundant in vernix and required by the rapidly growing cells of the maturing gut.[22] Vernix also contains compounds known as *surfactants*, which are important for keeping the airway sterile when the baby begins to breathe after birth. In its reported role as a promoter of wound healing, vernix that passes to the mother's perineum during birth would hasten healing of this region following episiotomy or any other nicks and abrasions that occur.[23] It also serves as a skin cleanser, which is important for both mother and baby at the time of birth.

In summary, this amazing multifunctional substance seems ideally suited for enhancing health and survival of newborn infants and postpartum mothers. An evolutionary perspective argues that something this complex and this useful (skin cleanser, moisturizer, anti-infective, and anti-oxidant) should probably be utilized to its maximum advantage rather than washed away with other seemingly noxious birth fluids. This would be especially important in health-poor countries where technological interventions are limited. The World Health Organization recommends that bathing infants should be delayed until a few hours after birth to minimize heat loss,[24] and certainly other important functions of vernix would be compromised by bathing.[25] The anti-infective properties of vernix are equally important in hospital births where it may protect babies from hospital-acquired infections.

Mothers Soothe and Calm Their Babies Immediately after Birth

Many of the things mothers do within minutes after birth serve to soothe the newborn infant, who often shows signs of stress from the experience of labor and delivery. Conserving energy during this time is important, so any behavior that reduces crying and other unnecessary energy expenditures would be advantageous. One behavior that has been observed across cultures is the tendency to hold the infant on the left side of the body, regardless of maternal handedness. More than 50 years ago, child psychologist Lee Salk reported that most mothers held their babies on their left sides in the first few days postpartum and proposed that this practice exposed the infant to the mother's heartbeat (a familiar sound in utero), which, in addition to soothing, promoted weight gain early in infancy.[26] Furthermore, infants seem to prefer to look to the right, which suggests that mothers may be responding to subtle cues from the infants when they decide how to hold them.[27] In my observations of 100 mothers, almost three quarters pulled their infants to their left sides when the babies were first handed to them and more than three quarters held left for most of the first hour after birth.[28] I have even speculated that the tendency to hold infants on the left side, over the heart, may have contributed to the emergence of lateralization and right handedness in hominin evolution.

A newborn baby seems to be exquisitely attuned to the mother's face and eyes. This is not surprising, considering that visual communication is one of the hallmarks of primates. Unlike truly altricial mammals, human infants' eyes are open at birth and they appear to be able to focus on objects 10–20 inches from their faces—about the distance from the mother's breast to her eyes when the baby is breastfeeding.

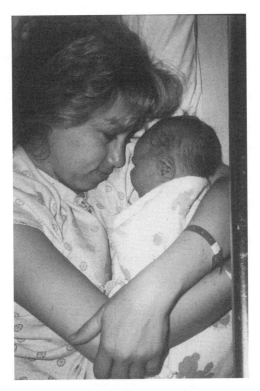

Figure 6-1 Mother looking *en face* at her newborn infant. Used by permission from Anne and Phillip Johnston.

Observers note that mothers try particularly hard to place their own faces in a plane with their infants' faces, a position known as *en face*[29] (see Figure 6-1). When infants are awake and looking into their mothers' faces, they tend to remain quiet, as if engaged in the process of learning about the mother.

Mothers all over the world use their voices to soothe their infants and even have a special language known as "motherese."[30] Vocal interactions between mothers and infants of most animal species also serve to maintain proximity and facilitate individual recognition and nursing. Often the maternal vocalizations are high-pitched, complementing the fact that infants are more able to perceive sounds in the higher ranges. Pediatrician Berry Brazelton suggests that the human neonatal nervous system is better able to respond to the higher pitched female voice.[31] Other observers note that mothers seem to "instinctively" elevate the pitch of their voices when talking to their infants, often switching pitch mid-sentence when they move from looking at adults to looking at their infants.[32]

An evolutionary perspective on left-side holding of infants, *en face* gazing, voice pitch elevation, and patterns of touch suggests that they may be the result of natural selection for behaviors that soothed infants, contributed to energy conservation and weight gain, and facilitated mother-infant bonding. The strong bond and associated behaviors also contribute to successful breastfeeding, surely one of the most important factors influencing infant survival in the past and in much of the world today. If these maternal behaviors in the first hour after birth are indeed part of the behavioral repertoire of the human species, it is not surprising that many women today report frustration when they are not able to be with their infants continuously from the moment of birth onward. Certainly if a challenge to the health of the mother or infant occurs at birth, separation may be warranted. It is difficult to bond to a dead baby, of course, so when reasons of health and survival lead to separation of mother and baby very few people would question that decision. But it seems obvious that mothers and babies benefit for a myriad of reasons from being together after birth when things are going well, so modifying hospital and clinic practices to ensure the togetherness that promotes optimal health and well-being seems appropriate.

Is the First Hour a Critical Period for Bonding?

Despite all the wonderful things that go on in the first hour after birth that promote mother-infant bonding, it is not in any way *critical* for initiating this process. Very strong and successful bonds form even when mothers and infants are separated for hours or days following surgical deliveries, premature births, infant distress, and other complications related to birth. Adoptive mothers and infants form strong bonds without the experience of labor and delivery (for the mother). Parents who are strongly motivated to form an attachment to their infants can overcome almost any imaginable obstacle and are far from dependent on biological mechanisms to do this.

But it may have been a critical period for bonding in the past, one that benefited from all of the behaviors that mothers and infants exhibit in their interaction during the first hour or so following birth. Consider that the stroking and massaging that most mothers do with their newborn infants in the first hour after birth serve to stimulate breathing, provide warmth, and rub the vernix (with its antibiotic properties) into the skin. By holding the infant over her heart on the left side of the body and talking to him in a high-pitched voice, a mother soothes and quiets him, thus enabling him to conserve energy at a vulnerable time. In the ancestral past, these behaviors probably greatly enhanced infant survival during the very vulnerable period right after birth. The seemingly inborn urge in the infant to find the mother's breast may have been one of the most important ways for initiating the hormonal cascade that stimulated contractions following birth, leading to the expulsion of the placenta and shrinkage of the uterus to prevent postpartum hemorrhage. In this way, the infant helped to save its mother's life, thus ensuring that he retained his source of food, comfort, and very survival. Thus, I have argued that infants remaining with their mothers in the first hour after birth may well have been critical for survival in the past.[33] The biological and behavioral mechanisms that are seen immediately after birth are the products of millions of years of selection to ensure survival of mothers and infants during what is probably the most vulnerable period in one's lifetime.

Baby Blues and Postpartum Depression

Not every woman who has a baby spends the first hours and days after birth in a state of oxytocin-induced bliss gazing into the eyes of her newborn infant. Perhaps equally often, though it seems to vary across populations, women experience the "baby blues" that may lead to postpartum depression (PPD), characterized by a loss of interest in virtually all activities.[34] Both cultural and biological factors have been implicated in postpartum negative mood. For example, even within the United States, there is ethnic variation in incidence, with Hispanic women reporting the lowest rates and Native Americans the highest.[35] Although anthropologists report a great deal of variation in postpartum negative mood and suggest that cultural factors have a greater impact than biology, at least one group of scholars found that unhappiness following birth was reported in all populations they surveyed,[36] suggesting that the potential for low mood after birth may be there for all women, but that there is extensive variation in how and whether it is manifested.

Social factors are often implicated in postpartum negative mood; women who report low social support and marriage problems are particularly at risk.[37] Difficulties during pregnancy, labor, and delivery, and infants who have health problems are also associated with a higher incidence of postpartum negative mood, and the need for social support is even greater in such circumstances. All of this makes sense from an evolutionary perspective because, as has been pointed out frequently, women not only benefit from social support in the early months of their infants' lives but the survival of their infants and their own reproductive success is greatly enhanced by this support, especially when provided by the father of the child or other relatives, whose own reproductive success benefits when they support the child and mother. When postpartum negative mood is associated with poor health of the infant, it may be part of the "cut your losses" phenomenon that characterized parenting in the past, sad as that may seem.[38] This is not to say that it is acceptable today to abandon or lower investment in a sickly child but just to remind us that circumstances surrounding parenting in the past were likely very different from those we face now, and understanding these feelings may lead to workable solutions to contemporary problems.

Anthropologist Ed Hagen has suggested that in many cases, postpartum negative mood may elicit sympathy or assistance from relatives (including the father of the child), adding support to the view that at low levels, negative mood may be advantageous.[39] He proposes that in the past, if the mother gave signs of abandoning her child, those with genes invested in that child would have been more likely to begin contributing time and resources to its care. Thus, she could use negative mood to "negotiate" assistance from members of her social group. Of interest with regard to the proposal that one of the legacies that pregnant women take with them into labor and delivery is a need for emotional support, women who have supportive companions (sometimes referred to as doulas) with them at the time of birth are less likely to report postpartum negative mood.[40]

Negative postpartum feelings can range from mild to severe, and they provide another example of ways in which we can view health challenges as defects or defenses. According to Randy Nesse, some negative feelings may have adaptive value, unless they reach a level at which the person is completely debilitated or unable to function.[41] A mild state of blueness after birth may lead a mother to focus all of her thoughts and energy on the task most immediate and, for fitness, most important: caring for her infant. While she is feeling blue, she may not be interested in going out shopping or to social engagements or back to work. These feelings may not be helpful today, but in the past their expression in new mothers may have resulted in babies that received more focused attention from their mothers and thus were more likely to survive. Clearly when the sadness reaches levels that prevent her engaging with her infant at all or compromise her own health, they become maladaptive and warrant treatment.

A practice that is widespread across cultures is seclusion of mothers and infants for a period of time after birth, ranging from a few to 40 days. This practice probably affords the mother time to recover from labor and delivery, time to initiate and further develop successful breastfeeding, and time to bond with her infant. Related to this idea of lower activity levels in women after birth contributing to better infant care, anthropologist Barbara Piperata reports a cultural practice of food restriction for women in the Amazon, known as *resguardo*, that lasts for 40–41 days and is associated with reduced overall energy expenditure.[42] Postpartum restrictions on mothers' activities are common worldwide and seem to function to ensure health of the mother and infant who are vulnerable at this time. Perhaps mildly low mood after birth in Western societies serves the same function, especially where cultural practices have not been preserved to sustain the woman at this time.

Another evolutionary take on postpartum low mood is related to the argument that in the past, if a baby was born under conditions in which survival was unlikely (if it had defects or was sickly, or if there was no one to help the mother and infant), it may have been advantageous to abandon the infant soon after birth. Low postpartum mood may have interfered with bonding at this stage and enabled the mother to act on what may have been in the best interests of her long-term fitness.[43] Related to this argument is the proposal that the high levels of prolactin associated with lactation in the first few weeks following birth contribute to heightened feelings of hostility toward others that may, in the past, have enabled women to be more protective of their new infants once they decided they were worth investing in. Sarah Hrdy calls this "lactational aggression" and notes that it is even manifested in mild negative feelings toward spouses in the few months following birth.[44]

Despite arguments that there may be advantages to mild low mood following birth, when postpartum depression becomes moderate to severe, it is likely to have negative effects on the mother-infant relationship that may extend beyond the first year of life.[45] It can interfere with breastfeeding,[46] which we will see is one route to lifelong compromised health for both the mother and infant. All other things being equal today, it is probably not particularly helpful for mothers to feel blue after birth, but in the past women who had mildly low mood may have been able to provide better mothering.

Inflammation, Hormones, and Baby Blues

A search for underlying biological factors that predispose or are protective against postpartum low mood and depression has yielded conflicting results, but there is little doubt that these factors exist and, with psychosocial and environmental factors, influence risk for postpartum low mood. Childbirth has a direct effect on the immune response and on the hormones that regulate stress response and there is increasing evidence that many physical and psychological ills that humans face are due to infection and inflammation.[47] In support of the proposal of an infectious link to postpartum low mood, some researchers found higher inflammatory response in women who had postpartum blues.[48] Thus, anything that can reduce inflammation can, presumably, reduce the chances of experiencing postpartum depression. Women who breastfeed are less likely to suffer from depression, suggesting another possible connection to infectious agents. The probable mechanism linking these is that women who have recently given birth are more susceptible to inflammation due to a number of factors, including hormonal changes, pain, sleep disturbances, and other stressors. All of these can increase the likelihood of depression, but breastfeeding (unless it causes pain and stress itself) can attenuate it by countering stress and thus the inflammatory effects.[49] Depressed mothers often stop breastfeeding, a decision that may perpetuate the depression. Because breastfeeding is so ubiquitous in traditional cultures, this may partially explain the low reported incidence of postpartum blues in those cultures.

Kathleen Kendall-Tackett, a health psychologist, claims that inflammation is not just one of the factors that influence postpartum mood, but is *the* factor.[50] She argues that rather than acting directly on mood, all other variables that seem to be related to postpartum feelings (pain, sleeplessness, trauma, lack of social support, marital difficulties, infant illness, low income, a history of trauma) act to increase inflammation which, in turn, increases risk for depression. Developing ways to deal with the inflammatory response to these stressors may be the best hope for intervention efforts and prevention of postpartum low mood and depression.[51] Getting a good night's sleep may be difficult for new mothers, but it is probably one of the best ways of ameliorating postpartum blues, just as it is for many other physical and psychological ailments.[52] Add breastfeeding to the prescription and postpartum depression may be avoided by most new mothers. Kendall-Tackett also suggests that anything a woman can do to reduce inflammation will also improve postpartum mood. This includes consuming anti-inflammatory agents such as long-chain omega-3 fatty acids.[53]

Unfortunately, because depression often leads to sleep deprivation, there is a vicious cycle that is difficult to overcome in the early postpartum days. It is the same with pain: pain increases stress and inflammatory response, which in turn, lowers the threshold for pain and leads to low mood. Almost all women report some level of pain in the postpartum period. Interestingly, anthropologists Jim McKenna and Thom McDade report that what may be considered to be disrupted sleep seen in women who co-sleep with their infants does not seem to be related to increased pain, suggesting that co-sleeping may be protective against low mood.[54] Women who sleep with their infants also tend to breastfeed and they do not usually have to get up in the night to feed, so their overall sleep may be better than that of a mother who arises every hour or two to breastfeed her infant.[55]

Finally, I return to the evidence that hormones of pregnancy reach much higher levels in women in health-rich nations than they do in women in health-poor populations. Thus, the changes that occur at the end of pregnancy and the withdrawal of high levels of estrogen and progesterone may have more profound impacts on health-rich women and thus lead to low mood. It seems that we can blame more and more of the ill effects of modern civilization on factors that increase lifetime exposure to high levels of reproductive hormones.

If those of us who work in the field of evolutionary medicine made recommendations, here is what we might say mothers should do to reduce the chances of developing more serious mood disorders of the postpartum period: (1) prepare for childbirth so that its physical and emotional stressfulness is reduced, especially by having a supportive companion with the mother at delivery; (2) eat plenty of omega-3 fatty acids throughout pregnancy and postpartum; (3) breastfeed the infant for at least a year; (4) get as much sleep as possible, even if it means sleeping with the infant (in a safe manner) so as to minimize the disruption caused by breastfeeding; (5) reduce stresses from as many sources as possible, even if it means some form of (nondrug) therapy to deal with ones that cannot be controlled; (6) get plenty of exercise, because it counteracts both inflammation and stress (but do not *start* doing vigorous exercise in pregnancy if exercise has not been routine); (7) keep potentially stressful infant crying to a minimum by breastfeeding frequently and responding to cries as quickly as possible. Even if none of these works to prevent baby blues, they enhance overall health of mothers and infants, so I have no hesitation in proffering them.

I have demonstrated that a lot of interesting and important things go on in the first hour after birth and that in the evolutionary past, mother-infant contact during this time was probably critical to infant survival. Given what is going on in the first year or two of life, it may seem strange that an entire chapter has been given over to what is essentially only a single hour. Perhaps I have paid too much attention to this tiny bit of time, but as noted earlier, it is probably one of the most significant hours in a person's life, and what happens in that time can potentially have a great deal of impact on how that life unfolds. It certainly had great importance in the evolutionary past. In the next chapter, I move beyond the immediate postpartum period to the first few months of life when infant growth predominates, including continued fat deposition and brain maturation. All of this happens under the watchful eyes of mothers and other caretakers, but the most crucial component is breastfeeding.

7

Women Are Defined by
Their Breasts

This is a provocative chapter title, but the entire class of animals to which humans belong is defined by the fact that they nurse their young with milk secreted from mammary glands. Like it or not, human females cannot escape the fact that much of their biology, including their breasts, is "designed" for producing and nourishing children. I suppose that to some readers it does not make sense for me to put a chapter on early infancy and childhood in a book about women's health, but for much of human history, infant health *was* maternal health, because so much of most women's time and energy was devoted to caring for the children in which they had already invested heavily by gestating them. The mammalian universal characteristic of lactation also has an effect on a woman's health, not the least because it is so costly. As a reminder of how costly, consider that nine months of gestation requires more than 340,000 calories and the equivalent time lactating costs more than 670,000 calories.

Biology of Lactation

Lactation, the process by which milk is produced and secreted, occurs in three stages: development of the mammary glands, production of milk, and nursing. Development of the mammary glands, which takes place during pregnancy, involves hormones such as prolactin, progesterone, and estrogens that prepare the breasts for nursing. Of course, breast and mammary gland development actually begins far earlier, in utero and at puberty, but undergoes full development in preparation for breastfeeding during pregnancy. While a woman is pregnant, milk secretion is inhibited by progesterone and estrogen. Production of milk does not begin until two to three days after birth; before that time, the breasts secrete *colostrum*. Colostrum is lower in calories and fat than mature milk, but it is more than twice as high in protein. More important, colostrum is high in immune factors that provide protection for the vulnerable newborn infant. Milk production can be delayed in women for a number of reasons, including retained placenta, caesarean section, diabetes, and extreme stress during birth.[1]

When breastfeeding begins, the hormones prolactin and oxytocin play important roles, but the real key is suckling—in other words, in the absence of an infant who suckles, milk production would cease soon after birth. (Because of the importance of suckling, women who give birth to premature infants who are not strong enough or developed enough to suckle are often advised to use breast pumps to mimic the infant suckling if they intend to breastfeed.) When the infant does suckle, a signal is sent to the hypothalamus in the mother's brain, which directs the secretion of dopamine and oxytocin, and inhibits secretion of hormones that would otherwise lead to ovulation (see Figure 7-1). Dopamine moves down one pathway through the anterior pituitary to increase prolactin, which, in turn, acts on the breast to maintain milk production if suckling continues. Oxytocin is released from the posterior pituitary and travels to the breast where it directs milk ejection (let down) when the infant suckles. As noted, the key to all of this interrelated functioning is infant suckling. Nursing is a supply and demand phenomenon: the more the baby suckles, the more milk is produced. For this reason, suckling twins or triplets does not pose a problem for most women. On the other hand, problems do arise when the infant is supplemented with other liquids, reducing suckling frequency and intensity, thus reducing milk production. There are

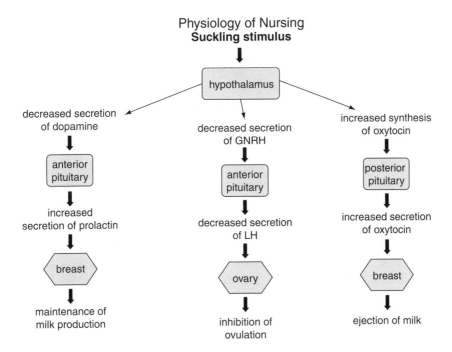

Figure 7-1 Hormones of breastfeeding.

probably very few women who have real physiological problems with milk production, but a much higher number stop breastfeeding because of "insufficient milk" production, primarily due to behaviors such as supplementation and the resulting decreased suckling.

Oxytocin is closely tied to breastfeeding, playing a role in let-down and in inducing a woman to feed her infant. In some cases, women experience let-down when they hear their (or any) infant cry, or even when they think about the infant. This demonstrates the important role of psychological and emotional processes in successful breastfeeding and points to ways in which psychosocial stress can inhibit let-down and lead to problems with breastfeeding.[2]

As noted earlier, breastfeeding has an effect on fecundity through inhibition of hormones responsible for ovulation (LH and FSH). Again, the key seems to be the suckling stimulus. When infants are suckling frequently (several times a day and throughout the night) and are not supplemented, ovulation can be inhibited for several months. When suckling rate and frequency are lowered with supplementation, women often resume cycling and can become pregnant. This means that breastfeeding is not a reliable form of contraception for most women. When examined at the population level, however, the pattern is one of delayed birth intervals as duration of breastfeeding increases.

A Question about the Standards Used in Medical Research for Infant Growth and Development

There is little doubt that breastfed infants are healthier, even though standards for measuring that health have only recently been developed. When we read about recommendations to support "normal growth and development of infants" it may not be clear what "normal" means in this context. Normal by whose standards? Most likely these are standards based on well-nourished populations (commonly European and Euro-American) living in healthful environments with access to resources including medical care. Furthermore, many were developed several decades ago based on infants who were not breastfed (or were only partially breastfed), lacked immunological maturity based on passively acquired resistance in breast milk, slept alone in cribs in separate rooms, and were allowed to cry until they fell asleep with exhaustion.[3] In contrast, it makes "evolutionary sense" that standards for "normal" infant growth and development should be based as closely as possible on what was the likely evolutionary environment of infancy, characterized by (1) full and exclusive breastfeeding until 6 months of age; (2) continued breastfeeding until about age 2 with supplementation gradually increasing from six months on; (3) infant-led ("on demand") breastfeeding; (4) no nutritional or emotional competition from siblings until at least age 3; (5) co-sleeping accompanied by breastfeeding during the night; (6) close physical and emotional contact with another person for most of the first year of life; and (7) immersion in dense social networks that provide support for parenting and child care. When infants who have a chance to grow up under these circumstances can be studied for developmental milestones, then we may be able to develop more evolutionarily appropriate standards of "normal" infant growth and development, and we can focus attention on deviations from those standards and their presumed causes.[4]

Most of what makes evolutionary sense as the "norm" is linked to long and intensive breastfeeding. In the 1960s in the United States, hospital births occurred with anesthesia, analgesics, and unfamiliar caretakers; routine separation of mother and infant at birth; infants housed in nurseries; and delayed or absent breastfeeding.[5] Today, although this scenario may still describe some births, more common are efforts to help mothers breastfeed, including doing so a few minutes following birth, in the belief that the sooner the baby begins to nurse, the more successful breastfeeding will be and the longer it will last. Unfortunately, this critical aspect of parenting is another one that often falls between the cracks of Western medicine, wherein one set of professionals treats mothers during pregnancy and birth and another takes care of babies after birth. There is not a profession that has as its focus the new mother who is breastfeeding an infant for the first time,[6] despite efforts to develop one.[7] The closest thing to a breastfeeding profession is organizations like La Leche League, which include volunteers who provide support to new nursing mothers. Simply knowing that "breast is best" is not sufficient in a culture that has devalued breastfeeding for two to three generations and where role models for successful breastfeeding have been rare until recently.

For several decades, the growth charts used by the World Health Organization and international health clinics were based on a sample of healthy European-Americans from a single community in the United States that relied on anthropometric data collected between 1929 and 1975 (known as the Fels Longitudinal Study). Most of the infants in the reference dataset were not breastfed or, if they were, not for very long. When these standards were used to assess breastfed infants, they almost always showed less-than-optimal growth for these infants from ages 8 to 12 months. In other words, the breastfed infants consistently were leaner and lighter at several measurement points than the reference standard based on formula-fed infants.[8] A few decades ago this was referred to as "growth faltering" and was used to argue against exclusive breastfeeding beyond three months.[9] At the time of this 1979 study, most babies in health-rich countries were not breastfed so the primary concern was that breast milk was inadequate to meet the needs of babies in health-poor countries. The authors argued that proposals to have infants fed exclusively at the breast to avoid contamination and infections in unhygienic environments were "unjustified,"[a] given the risk of falling below the growth charts that we now know to be based on inappropriate standards. Ironically, this comment about "unjustified" recommendations came from an article proposing a Swedish code of ethics for marketing of infant foods.

In hindsight, it should seem odd that the type of food that human infants have depended on for millions of years would be deemed inadequate in any way, especially when assessed across populations. To many breastfeeding advocates, it seems logical that healthy breastfed babies would serve as the reference standard rather than babies fed with artificial substitutes. In an article entitled "Breast Is No Longer Best: Promoting Normal Infant Feeding," health researchers Nina Berry and Karleen Gribble argue that breastfeeding should no longer be seen as the *optimal* way to feed infants but rather should be viewed as the *normal* way.[10] Then the alternatives (like formula feeding) could be evaluated from the more appropriate baseline.

Recognizing the problem with comparing breastfed babies to growth charts based on formula-fed infants, the World Health Organization constructed a new chart and recommended that it be used not only as a guideline but in a proscriptive manner, as an indication of how infants *should* grow when their mothers follow healthful practices by breastfeeding them. The goal was "to describe the growth of healthy children,"[11] leaving little doubt that there was any better way to feed babies. When the earlier growth charts were used, overweight infants and children were judged to be normal, contributing to the explosion of obesity worldwide.[12]

Another example of how formula feeding has affected a medical view of "normal" has to do with the thymus, an organ that plays a major role in immune function. What was regarded as the normal size of a healthy thymus in infants was based on studies of babies who were formula fed during the years when breastfeeding was uncommon in the United States. In one study of the size of the thymus in healthy infants at birth and age 4 months who had various feeding routines, babies who were breastfed had much larger thymuses than babies who were partially or fully formula fed.[13] The size was still larger in infants who were breastfed at 10 months compared with those who stopped breastfeeding by that time. In fact, the size was directly correlated with the number of feeds per day.[14] The researchers conclude that immune factors in breast milk caused an increase in thymus size. The thymus and its immune function tend to increase from birth until puberty, after which time it steadily decreases in size. According to some researchers, the thymus of breastfed babies should be regarded as the normal size and the non-breastfed infant thymus as suppressed in size rather than the larger thymus being regarded as abnormal. In general, as Thom McDade has pointed out, "normal immune function" for humans has long been based on studies of overnourished and underinfected populations.[15]

Another concern is the question about what is the "normal" length of time for breastfeeding. Not surprisingly, the answer to the question is highly culturally contextualized. Normal to a South African San or Australian aborigine woman may be three to four years. Normal to an American woman may be fewer than six months. The American Academy of Pediatrics (AAP) recommends exclusive breast-feeding for six months and breastfeeding with supplementation for a year or more, depending on the wishes of mother and infant.[16] Because fewer than 30% of American women breastfeed for as long as six months, breastfeeding for a year or more is likely to be seen as abnormal or at least unusual in the United States. I will discuss duration of breastfeeding later in this chapter.

Finally, respiratory, gastrointestinal, and middle-ear infections are so common in early infancy that they are seen as "normal" and expected by parents and health care practitioners. But to a great extent, this is because they are so much more frequent in infants who are not breastfed. Frequent crying from hunger is another "normal" facet of infancy, providing yet another example of how "the normalization of formula feeding redefined 'normal' infant and child health."[17] Later in this chapter I will discuss developmental milestones for infants that are often portrayed as "normal" but which actually vary in surprising ways.

Why Breastfeeding Enhances Health for Infants and Mothers

As noted, many of the arguments I present here pertain to health consequences that have resulted from natural selection operating to increase reproductive success. Because successful breastfeeding was so important to infant survival in the past and in many parts of the world today, it was and still is also important to maternal reproductive success. For these reasons, it makes sense to discuss some of the ways in which breastfeeding ensures survival and health even today when pretty good substitutes for mothers' milk are available. Not perfect or equivalent, admittedly, but just "pretty good." We will see that beyond this, breastfeeding seems to contribute to positive health outcomes for mothers in addition to the obvious ones that infants gain. In other words, breastfeeding is good for moms, too.

The costs of lactation are great for mammalian mothers, but survival rate is higher for their young than for the offspring of nonmammalian species who have to begin finding food on their own soon after they are born. The newborn mammalian infant is so dependent on the mother for nourishment that it is sometimes referred to as a parasite. Although the usual definition of a parasite is an organism that depends "on something else for existence or support without making a useful or adequate return,"[18] we will see that mothers get a lot in return for breastfeeding their young. But the daily caloric intake required of human mothers who are lactating is higher than that required during pregnancy, so it seems that anything that could relieve this burden, such as using milk substitutes, would be a "good thing" if the substitute is equal to the milk mothers produce. Of course, the problem is that true equivalency has so far proved elusive despite the best scientific efforts.

Although the nutritional qualities of human milk have been touted in hundreds of articles in both the academic and popular press, it bears repeating here that the composition of milk is finely tuned to growth and development requirements of all mammals and that substituting what mothers produce with the milk of another species or carefully designed artificial substitutes (formula) will in almost all cases have unintended, sometimes negative, consequences. Furthermore, milk composition is tightly correlated with nursing patterns for each species and tampering with those patterns may also have negative consequences. For example, lions, which need to spend several hours away from their infants while they hunt, have milk that is rich in nutrients and calories so that the infants can be satiated with feedings spaced every six to eight hours. More than one third of their milk is comprised of fats, carbohydrates, and proteins. In contrast, marsupials like the red kangaroos who are in constant contact with their young (in fact the embryonic young of some species become fused to the teat in the pouch following uterine delivery), which nurse more or less continuously, have milk that is comprised of less than 15% nutrients (85% is water) (see Figure 7-2).

Primates, including monkeys, apes, and humans, also have milk that is low in nutrients, which is evidence that the "natural" nursing pattern is one involving almost continuous contact and frequent nursing. In the 1950s, pediatricians in the United States recommended feeding infants every four hours, approximating a pattern seen in rabbits whose milk is comprised of 28% nutrients, more than twice the density of human milk. To protect themselves and their young from predators, rabbits need to nurse quickly and then remain away from their infants for about four hours at a time. Humans are not rabbits and their milk is not rich enough for their infants to survive with rabbit-like nursing patterns. An evolutionary view points to behavioral ecology of other mammals, especially closely related apes and monkeys, for evidence of how frequently human babies should be nursed.

Milk composition and nursing behaviors evolved together and even the micronutrients are finely tuned to infant requirements. Fat composition increases from the beginning to the end of a nursing bout, something that formula simply cannot mimic. Furthermore, milk composition changes as the infant grows, reflecting changing nutritional requirements. One view that derives from an understanding of milk composition is that women today have inherited from their hominin ancestors milk that is more than 85% water, requiring highly frequent feeds (four to five times an hour in the early months) and thus more or less constant contact with their

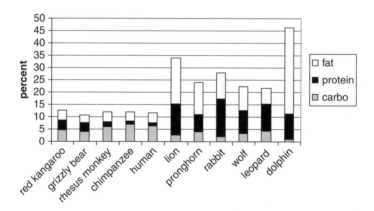

Figure 7-2 Carbohydrate, protein, and fat content of milk of selected mammals.

infants. Related to this are physiological mechanisms to suppress ovulation, emotional mechanisms to facilitate optimal psychosocial development in the infant, and social mechanisms to ensure support for mothers and infants. As I have previously argued, "The social and demographic characteristics of our species . . . have been shaped largely by this pattern of delivering relatively helpless infants, requiring frequent feedings, and greatly intensified parental care." [19]

The Case for Breastfeeding Infants

Of course, mother's milk has many benefits in addition to nutrients. Among the constituents are proteins (enzymes, growth factors, hormones, and immune factors), carbohydrates (lactose), lipids (fat soluble vitamins and fatty acids), water soluble vitamins (11 of them), minerals (20 of them), and other cells that are involved in immune function (18). Particularly important for brain development are the long-chain polyunsaturated fatty acids (LCPUFAs). Women who deliver prematurely have milk higher in these fatty acids than women who deliver at term, suggesting that their premature infants need the higher amounts for their development.[20] Infants who are born at full term store up these fatty acids during the last couple of weeks of gestation, a period not available to prematurely delivered infants. It should be noted that fats are not easily absorbed by young infants but are aided by enzymes unique to human milk, which explains why fats in the milk of another species or in formula are not well digested by human infants.

Primate milk is unusual in the high amount of carbohydrates, most significantly the milk sugar lactose, which is easily converted to glucose during digestion. Glucose is important for species with rapid development of exceedingly hungry brains, a phenomenon characteristic of primates, especially humans.[21] Recent research suggests that cognitive development is better in infants who were breastfed, giving them a head start that extends even into the teens. The LCPUFAs in human milk play important roles in brain development and associated cognition. A review of 20 carefully controlled studies of cognitive development that were conducted between 1966 and 1996 found that the benefits were sustained from early infancy to adolescence and that as duration of breastfeeding increased, so did cognitive development.[22] Babies who were born with low birth weights or prematurely benefited even more from breastfeeding. In addition to general cognitive development, breastfed infants show enhanced visual skills, earlier motor maturation, and fewer behavioral problems. This may be the reason behind better cognitive development in breastfed premature infants who otherwise would not get sufficient fatty acids from the mother in the late stages of pregnancy.

There is controversy about the link between breastfeeding and intelligence, so not surprisingly there are contradictory reports of the effects of breastfeeding on intelligence. One study examined the association and concluded that it was the mothers' IQ scores that were responsible for the assumed effects; women with higher IQs were more likely to breastfeed their infants and, since IQ is highly heritable, they had infants with higher IQs, perhaps irrespective of breastfeeding.[23] Other scholars have found that the home environment, including amount of intellectual stimulation and emotional support provided by the parents, have a greater impact on cognitive development than breastfeeding.[24] Although it still seems likely that there is a benefit to brain development provided in breast milk, especially for premature infants and considering the fatty acids present, these more recent findings should give comfort to women who are unable to breastfeed their children.[25]

There is increasing evidence that breastfeeding may be protective against several adult-onset diseases and disorders including type 2 diabetes, obesity, hypertension, and various cancers. For example, breastfeeding exclusively for at least two months appears to be associated with a lowered incidence of type 2 diabetes in Pima Indians, a population that has the highest incidence of the disorder in the world.[26] Breastfeeding is also associated with a decreased risk of type 2 diabetes among Native Canadian children.[27] Mechanisms proposed include hormonal differences and the more optimal weights that breastfed infants have because they consume less during feeds. In support of this, some researchers found that as breastfeeding duration increases, risk of being overweight decreases.[28] This makes sense when we consider that human milk is tailored for optimal growth of human infants. If these relationships hold up in other populations,[29] it provides hope for lowering the worldwide rates of obesity and diabetes by increasing breastfeeding rates, but it is also a bit unsettling to think that lifelong weight levels may be established in the first six months of life.

In another example of early life influences on later adult health, Indian infants who were breastfed had better lipid profiles (better HDL to LDL ratios) in the first six months of life than did those who were fed according to a mixed strategy (breast milk supplemented with formula or animal milk), suggesting a mechanism for early childhood influences on the development of coronary heart disease.[30] Additionally, a 65-year follow-up of a group of British subjects indicated a reduced risk of atherosclerosis in those who were breastfed.[31] Finally, in a review of 15 studies and more than 17,000 participants, having been breastfed as an infant was associated with small reductions in both systolic and diastolic blood pressure.[32] The authors argue that even the small difference they found could "reduce the prevalence of hypertension by up to 17 percent, the number of coronary heart disease events by 6 percent, and strokes and transient ischemic attacks by 15 percent."[33] I should note that many studies have reached different conclusions from the ones I present here, but for women who are seeking reinforcement for their intensions to breastfeed, there is plenty of evidence of positive benefits.

Table 7-1 compares nutritional requirements for a mother of selected nutrients for the nine months of pregnancy and the first nine months of infancy. From this table it is easy to see that nine months of lactation "cost" more than the entire pregnancy, especially in calories (almost twice as expensive) and vitamin A. In fact, the only nutrients in this table that are required in higher amounts during pregnancy are folate and iron. Human milk is limited in iron availability and much of the

Table 7-1 Nutritional requirements during pregnancy and lactation for selected nutrients

Nutritional requirement	Pregnancy (9 months)	Lactation (9 months)	Especially important for
Energy (kJ)	340,200	676,620	Brain development
Protein			
Vitamin C (mg)	18,900	19,800	
Thiamin (mg)	389	405	
Riboflavin (mg)	378	432	
Folate (mug DFE)	162,000	135,000	
Vitamin A (mug RE)	234,000	351,000	
Vitamin D (mug)	1,350	1,350	Bone mineralization
Vitamin E (mg alpha-TE)	2,700	3,240	
Calcium (mg)	270,000	270,000	Skeletal development
Phosphorus (mg)	324,000	324,000	Skeletal development
Iron (mg)	8,100	4,050	Immune function
Iodine (mug)	47,250	54,000	Thyroid function; brain development

DFE = dietary folate equivalents; RE = retinol equivalents; TE = alpha-tocopherol equivalents.
From Picciano, 2001.

iron needed in the first few months of life comes from stored iron acquired during the last few months of pregnancy, which is why adequate iron during pregnancy is so important for both mother and infant. As noted, iron deficiency is the most common nutritional deficiency of pregnancy worldwide, including in health-rich nations, and the low amount in milk after six months is the primary reason for advising supplementation from that point on.

Colostrum and Milk Provide Immune Protection

Like all mammals, newborn human infants are born with poorly developed immune systems and, initially, their major defenses are those derived from the mother while they were in utero. But further development of immune function and protection from infectious agents in the days and weeks before the immune system matures depends on defenses derived from mothers' milk.[34] It may not seem sensible from an evolutionary perspective that a young infant, in whom so much energy has been invested for the previous nine months, is so vulnerable to the thousands of potentially hazardous microorganisms to which he is exposed at birth. But that vulnerability is reduced significantly if infants begin to nurse soon after birth, availing themselves of the colostrum, which is notably high in immune factors. These immune factors are simply not available in any other milk or milk substitute that may be used to feed an infant. Furthermore, maternal immunity to some pathogenic conditions in her environment (such as tuberculosis and schistosomiasis) may also be transferred to infants via milk so that the infant is protected from the specific microorganisms of the environment into which he is born.[35] Although some immune bodies may be highly specific to a particular mother and her infant, most are highly conserved in evolution and are found in all mammals, including monotremes (like the echidna and platypus) who secrete them through their skin (for the infants to lick) rather than via teats.

As with many other characteristics, the immune system of humans is most similar to that of chimpanzees, and it is more similar to that of apes and Old World monkeys than New World monkeys, confirming other lines of evidence of evolutionary relationships among primates.[36] To some extent, this reflects similarities in the relationship between mothers and infants in these species and the types of microorganisms to which young infants may be exposed. For example, calves and human infants have very different relationships with their mothers, and the immune composition of the milk reflects this. Calves are able to move independently from their mothers, thus becoming exposed to microorganisms that may be slightly different from hers, whereas young primate infants tend to remain on the mothers' bodies for the first several days and weeks of life. Cow and human milk are strikingly different in immunologic composition (see Table 7-2), reflecting the different mother-infant relationships and exposure to microorganisms in the early postpartum period.

But this still begs the question of why newborn infants have such undeveloped immune systems at birth. One possible explanation is that while in utero the infant was sufficiently protected from pathogens by the mother's defenses so that energy that would be required to develop and maintain the infant's own immune system could be used for growth and development of other systems. Second, as noted in Chapter 3, the fetus's immune function is delayed so as to avoid immunologic rejection by the mother. In the evolutionary past, vulnerability was reduced by passing antibodies through nursing soon after birth, so that the negative consequences of immature immune function were rarely manifested. Furthermore, it is

Table 7-2 A comparison of relative amounts of selected antimicrobial proteins in human and cow milk

Immunological bodies	Human	Cow
Secretory IgA	++++	+
IgG	+	++++
Lactoferrin	++++	+
Lactoperoxidase	±	++++
Lysozyme	++++	±

From Goldman, Chheda, and Garofalo, 1998.

probably cost effective to obtain immune proteins from the mother for as long as possible because of the costs of developing and maintaining the infant's immune system. Saved nutrients can be used for the developing brain and other systems crucial for survival that cannot be "parasitized" from the mother. By the time the infant begins to explore on his own and leaves the mother's body for minutes at a time, his own immune system has begun to mature.

Breastfed babies have lower incidence of diarrhea and other gastrointestinal diseases and both lower and upper respiratory illnesses.[37] These may seem like minor annoyances to people in developed nations where babies do not become hospitalized or die from them, but two of the leading causes of death of children under 5 in the world today are diarrhea and acute respiratory illnesses.[38] A huge number of these deaths could be avoided if mothers breastfed their children (with supplementation beyond six months) for the first two to three years of their lives. Furthermore, almost all of the leading causes of deaths of children worldwide are underlain by mal- and undernutrition, other factors that could be improved with prolonged breastfeeding. Well-nourished infants and children either do not get these diseases or are able to quickly recover from them.[39]

Breastfeeding appears to be protective against asthma, with a dose-response effect and better outcomes if the infant is breastfed more than nine months.[40] Skyrocketing asthma rates worldwide are related to obesity, respiratory infections, and breastfeeding. Breastfeeding appears to be the crucial link in these associations because children who are breastfed are less likely to be obese, to suffer respiratory infections, and to develop asthma.[41] Breastfed children are also less likely to be exposed to factors in other food sources that are believed to be linked to asthma.

Anthropologists Carol Worthman and Thom McDade have referred to the immune protection acquired by the infant from the mother via the placenta and her milk as an example of Lamarckian evolution (acquired characteristics) that gives babies a head start before their own immune capabilities (the result of natural selection, or Darwinian evolution) kick in.[42] Because breastfeeding practices are so heavily influenced by cultural norms, this means that culture has a huge impact on the development of immune function in infancy, right down to the size of the thymus. We are familiar with the concept that culture has a powerful influence on health, but the idea that it has a direct effect on immune function is somewhat new, and a bit sobering when we consider the lifelong consequences of a decision about breastfeeding that many women make without being fully informed.

Finally, there are qualities that breastfeeding offers beyond nutrients and immune protection. For example, when an infant suckles, gastrointestinal hormones are released that function to optimize energy intake and utilization so that growth is better than it would be if the baby were simply given the same calories via a catheter.[43] One of the hormones (cholecystokinin) is sleep-inducing, which conserves energy. In fact, suckling alone, even when no nutrients are obtained, enhances growth in premature infants.[44] Neurophysiologist Kerstin Uvnäs-Moberg suggests that pacifiers, in contrast to their negative press, actually serve a physiological function because they enable the baby to simulate the highly frequent suckling that characterized ancestral breastfeeding patterns. This is further evidence that modern alterations in breastfeeding behaviors have more than incidental consequences.

Breastfeeding and Excessive Infant Crying

Much of the current discussion about breastfeeding challenges the advice from pediatricians and child care experts of the 1950s. One of the recommendations that has been most controversial over the past several decades is the idea that babies should not be coddled when they cry and that the best strategy is to let them "cry it out" rather than indulging them by comforting them. Furthermore, infant crying is seen as routine and "normal" by parents today and is just part and parcel of having a baby in the house. An evolutionary perspective has a number of things to say about infant crying, most of which suggest that routine, prolonged crying by infants was probably relatively rare in the past.[45] Babies cry for three good reasons: when they are in pain, hungry, or alone. A mother who kept her infant with her all the time and provided opportunities for breastfeeding "on demand" probably did not have an infant who cried for hunger or loneliness, so pain would have been the only reason that a baby cried, and response to that cry of pain was likely immediate. Also notable is that nonhuman primate infants rarely cry, and they are also rarely away from their mothers.

Psychologist Nick Thompson and his colleagues have suggested that infant crying in the past may have served as a signal to others of hunger due to problems the mother was experiencing with her own lactation,[46] something that may have resulted, for example, if she had experienced emotional stress from fear or anxiety. They argue that in this case, an infant cry may have alerted other lactating women in the group that his needs were not being met, communicating to them that they should supplement by serving as "wet nurses." Although failure of lactation was probably rare in the past, in the event that it did occur, communication of insufficient milk by a hungry crying infant may have enhanced his chances of surviving. In their view, the cry was selected as a way of communicating hunger, not necessarily to the mother, but to other lactating women in the social group. Of course, if the infant suckled from other women, he was not suckling from his own mother, perhaps further compounding her problems with milk production.

There are several reasons that it would have been advantageous in the past for babies not to cry and for parents to keep crying to a minimum. A crying infant is a signal that a vulnerable social group (mothers and young children) is nearby, which may be attractive to predators. Persistent crying has a profound negative and irritating effect on other members of a social group, so keeping crying to a minimum may have reduced physical abuse directed toward the infant. (Persistent crying of the infant is often cited as a reason for child abuse or infanticide in the United States.)[47] Crying is energetically expensive and may be particularly taxing on an infant in the early days of life when maintaining energy stores is especially critical. Finally, in the past it would have been adaptive for infant crying to signal only pain or danger, so that members of the social group could respond immediately when that signal was given. Infants who cried routinely for relatively trivial and easily modifiable reasons (hunger or loneliness) may have been ignored in the past when a true cry of pain was given, just as the boy who cried "wolf" too many times was ignored when the real thing appeared.[48] Keeping infants with them at all times and breastfeeding when they were hungry probably enhanced reproductive success of ancestral hominin females and seems to contribute to maternal well-being today. Mothers today know that their infants are not usually in danger when they cry, but babies do not know that when they find themselves alone and hungry.[49]

Many of the emotional stresses that mothers cite today as contributing to ill health and depression following birth are understandable from an evolutionary perspective if they are trying to raise their infants in ways that their mothers, friends, and perhaps professionals advise, when the advice includes nursing every three or four hours or not at all, putting the baby in a crib in a separate bed or room from that used by the parents, and letting the baby cry until exhaustion. Babies who are treated this way often develop colic, a condition characterized by persistent, frequent, and unsoothable crying with no evidence of underlying disease or disorder. As is probably obvious, this behavior is exhausting and extremely stressful for parents, and among other responses, a mother often terminates breastfeeding because she thinks inadequate milk is the cause of her infant's stress.

From an evolutionary perspective, the colicky baby is crying because his expectations are mismatched with the caretaking responses of the mother and he is letting his dissatisfaction be known. In pediatrician Ron Barr's terms, colic is not something babies *have*; rather, it is something that they *do* in an effort to bring caretaking practices in line with their needs.[50] Babies who cried when they were left alone in the past survived if their parents responded appropriately and the behavior has been passed on to their descendants. But today, the persistent crying of colic makes parents' lives miserable. Furthermore, it reduces parental confidence and well-being, which, in turn, often has negative effects on the parent-child bond. In many cases, the solution to colic and prolonged crying is found in maternal intuition, which often leads women to respond to and hold their infants despite recommendations to avoid indulging them so they can "learn" to be independent. An evolutionary perspective suggests a much wider range of healthful and appropriate parenting behaviors than are typically found in today's childcare guidebooks. "Listen to your own intuition" could be a crude mantra for relieving many of the minor (and sometimes major) stresses of parenting.[51]

Excessive crying is also harmful to the health of infants.[52] It is energetically expensive and can be particularly problematic for infants in the first few weeks of life when respiratory and metabolic function are not well established. In particular, crying can compromise oxygen transport in the neonate in addition to being exhausting. In fact, it has been suggested that some newborn infants would cry even more than they do if they were physiologically able.[53] Excessive crying can also have a negative impact on gastrointestinal stress, especially in infants older than 3 months and especially if the crying interferes with their ability to breast-feed.[54] One proposal is that gastro-esophageal reflux disease (GORD or "acid reflux") in infants is due to the "misalignment of culture and biology"[55] and argues that biocultural factors influencing breastfeeding and sleeping patterns and parental responsiveness and behavior are the "causes" of this disorder in most cases. For example, almost half of the cases of GORD in infants between 3 and 12 months of age are due to a negative response to cow's milk. The "cure" for this problem from an evolutionary perspective is to treat babies as they expect to be treated, based on their evolutionary history. Clearly, the health of the mother (and other household members) will be improved as well if excessive crying is reduced.

A number of studies have reported that rocking movements serve to quiet a crying baby, as long as they are above 60 cycles per minute.[56] This occurs at a slow walking speed for humans, suggesting that the positive response to this speed may result from millions of years of being carried on the mother's hip or in a sling around her neck as she foraged for food. One suggestion is that infant development proceeds best when babies are carried and exposed to rhythmic movements of walking. Today's parents try to approximate that speed by placing an infant in a mechanical rocker or swing. As another example, some frustrated parents who want to quiet their colicky infants place them "on top of the washing machine during the agitation cycle."[57]

Noting that the characteristics of the infant cry change at age 3 months, anthropologist Joseph Soltis has another evolutionary take on crying and suggests that the "healthy" cry in the first few months is a signal to the parents that a baby is likely to survive and is worth investing in.[58] He focuses on acoustic properties of the early infant cry, suggesting that a cry indicating chronic illness (a weak or an excessively high- or low-pitched cry) will elicit different caretaking responses from parents than a strong, robust cry. In this view, neglect (or outright killing) may be one response to a weak infant cry as part of an evolved strategy to increase lifetime reproductive success (again, the "cut your losses" strategy for terminating invest-ment in offspring that are likely to die). Notably, cases of abuse and neglect in infants are often associated with chronic illness and the associated "abnormal crying."

Despite its apparent rarity in most human cultures and in other primates, infant crying has been proposed to have functions beyond signaling needs to and main-taining contact with parents and other caretakers. Primatologist Kim Bard attri-butes excessive infant crying in humans to delayed brain development, especially in areas affecting vision.[59] Related to this is another proposal that because brain development requires external stimulation, when a baby is not exposed to environ-mental stimulation, crying helps to maintain "ideal levels of brain activation." [60] Traditional foragers and human ancestors, because they carried their infants with them at all times, had infants who did not lack for stimulation and thus did not "need" to cry. Finally, paleoanthropologist Dean Falk proposes that infant crying plays a role in the evolution of language in our species, partially explaining its rarity in nonhuman primates.[61]

The Importance of Breastfeeding for Mothers

Beyond its importance in securing reproductive success, breastfeeding has numerous health benefits for the mother. Endorphins and oxytocin are released into her system when she breastfeeds—these feel-good opioids and hormones contribute to positive interactions with and feelings about the infant, which enhance the attachment process. They also help to speed her own recovery from labor and delivery and seem to have antistress effects.[62] Miriam Lobbok, a physician, has called breastfeeding a "preventive health measure for women" and the "final stage of labor."[63] As I have noted, breastfeeding soon after delivery is one of the best "natural" ways to protect against postpartum hemorrhage, and continuing to breastfeed helps the uterus return to its pre-pregnant state and the mother to her pre-pregnant weight.[64] In the births with which I was involved, the midwives always encouraged infant contact with the nipple because other tools for preventing postpartum hemorrhage were not usually available to them.

After birth, energy allocation continues to be a challenge for the mother who breastfeeds her child, as did virtually all women in the past. But, just as in pregnancy, her physiology is optimized so she can make the most of what is available. Oxytocin, in particular, which is released during suckling, aids in energy conservation so that women can continue to reproduce, even under conditions of scarcity.[65] One of the reasons women put on weight during pregnancy is that extra fat serves as a storage container from which they can draw calories while breastfeeding.[66] Other ways in which physiology is optimized are seen in improved ability to mobilize fat from thighs and buttocks and lowered heat generation, specifically in muscles.[67] Today, when excess fat may prove problematic for long-term health, breastfeeding regulates metabolism to convert calories from stored fats to energy for the infant rather than clogging arteries that may subsequently lead to increased risk of heart attack.[68] These are energy-saving processes that enable a woman to reduce her energy expenditures when she is lactating. Suckling also regulates the mother's gastrointestinal physiology and helps her conserve energy at a time when her caloric needs rise steeply.

As noted with regard to heart disease, a woman's decision about breastfeeding may have lifelong effects on her health. In the Nurses' Health Study (NHS), women who breastfed for at least two years of their lives were 19% less likely to suffer a heart attack than those who did not breastfeed at all, based on a study of more than 96,000 nurses between 1986 and 2002.[69] Increased length of lactation was also associated with a decreased incidence of type 2 diabetes.[70] Breastfeeding appears to reduce severity of anemia, bladder and other infections, and spinal and hip fracture after menopause.[71] This last finding is especially interesting in light of the short-term loss of bone calcium seen in breastfeeding women, as it appears that the long-term benefits are positive with regard to osteoporosis.

Premenopausal breast cancer and ovarian cancer risks are lower for women who breastfeed. The protective effect of breastfeeding for premenopausal breast cancer seems to increase as the number of months and years of breastfeeding increases: women who had more than six years of breastfeeding had only about one third the risk of breast cancer of women who never breastfed.[72] These findings seem to hold up for women in health-rich and health-poor nations, suggesting that the difference in breast cancer rates between these two levels of development may be explained in part by different rates and duration of breastfeeding.[73] For endometrial cancer, number of years of breastfeeding is associated with decreased incidence for women until menopause, after which time the protective effects do not seem to persist.[74] Among the factors influencing cancer rates may be the decreased number of ovulations that breastfeeding women experience in their lifetimes, as discussed in Chapter 2.

Of course, there is a cost to breastfeeding for the mother, and a woman who breastfeeds for too long may compromise not only her future reproductive success but also her own health through what is referred to as "maternal depletion." This is particularly problematic where access to adequate nutritional resources is limited. In these cases, a woman faces another trade-off between prolonging interbirth intervals, thus providing her time to recover from the stresses of pregnancy and lactation, and avoiding depletion of her own body through extended breast-feeding.[75] By this time, the benefits for the infant of pathogen protection and nutrients may also be diminishing. The point at which the costs of breastfeeding for the mother exceed benefits is highly variable across populations and depends on availability of resources, maternal workloads, pathogen exposure risks, and social support systems. For example, in health-rich nations where risk of exposure to pathogens is low, the point of trade-off may be lower than in environments where pathogens are rampant. As Thom McDade and Carol Worthman note, "Local cultural ecologies shape the parameters for these tradeoffs and, ultimately, determine the pattern of breastfeeding and infant and maternal outcome."[76] This perspective also argues against universal recommendations for breastfeeding and supplementation.

Why Would Anyone Not Breastfeed?

Table 7-3 summarizes the benefits of breastfeeding for infants based on a survey of the scientific literature, most of which has been published since 2000. The benefits are staggering and it is amazing that anyone who knew about them would choose to not breastfeed except under extreme circumstances. Even in health-rich nations, breastfeeding may be protective of infant health: recent studies show that infants breastfed for at least six months were 3.5 times less likely to be hospitalized for respiratory infections, 2 times less likely to have diarrhea, and 1.6 times less likely to

Table 7-3 Benefits of breastfeeding for infants and mothers

Proposed benefits of breastfeeding	Source
Larger thymus implicating higher level of immune function through increased production of lymphocytes	Hasselbalch et al., 1996, 1999
Bioactive Substances	
Immunoglobulins in milk, especially IgA, which protects from pathogens in immediate environment of mother and infant	Various; Hamosh, 2001 and references therein
Vitamins that scavenge oxygen radicals and are anti-inflammatory	Various; Hamosh, 2001 and references therein
Hormones that enhance gastrointestinal development in the infant	Various; Hamosh, 2001 and references therein
Lactoferrin and iron—anti-infective (e.g., against *E. coli* and *Shigella*)	Various; Hamosh, 2001 and references therein
Milk protein casein—protective against bacterial infection (e.g., *Streptococcus* and *Haemophilus*)	Various; Hamosh, 2001 and references therein
Promote gut maturation	Various; Hamosh, 2001 and references therein
Nucleotides that enhance intestinal repair after injury and may strengthen immune response to vaccine	Various; Hamosh, 2001 and references therein
Enzymes that enhance digestive ability	Various; Hamosh, 2001 and references therein
Hormones and growth factors to enhance growths and development	Various; Hamosh, 2001 and references therein
Prolactin—regulates neuroendocrine development in infants and later in life	Various; Hamosh, 2001 and references therein
Limits contact with potential contaminants in bottles, water, food	Heinig, 2001 and references therein
Enhanced vaccine response	Heinig, 2001 and references therein
Reduced incidence of diarrhea and other gastrointestinal illnesses and dehydration	Heinig, 2001 and references therein
Fewer respiratory illnesses	Heinig, 2001 and references therein
Reduced otitis media (middle ear infections), urinary tract infections	Heinig, 2001 and references therein
Reduced incidence of SIDS (sudden infant death syndrome) (Alm et al., 2002)	McKenna, Mosko, and Richard, 1999
Protective against chronic diseases like type 1 diabetes, celiac disease, inflammatory bowel disease, atopic disease, multiple sclerosis, obesity, Crohn's disease	Davis, 2001 and references therein

(continued)

Table 7-3 (Continued)

Proposed benefits of breastfeeding	Source
Lower mortality from atherosclerosis	Martin et al., 2005
Better lipid profiles in early infancy	Harit et al., 2008
Lowered asthma rates	Oddy et al., 2004; Dell and To, 2001; Oddy, 2004
Childhood lymphomas	Bener, Denic, and Galadari, 2001
Inhibits fertility of the mother, increasing birth spacing to more optimal intervals	Various

Higher cognitive development
    Better visual skills
    Earlier motor development
    Fewer behavioral problems late in life
	Anderson et al., 1999 and references therein (but see Der et al, 2006; Jacobson and Jacobson, 2006)

Lowered incidence of type two diabetes	Pettitt et al, 1997; Young et al, 2002, but see Simmons, 1997
Lowered risk of being overweight	Harder et al., 2005
Lower systolic and diastolic blood pressure later in life	Martin et al., 2005
Higher upward social mobility	Martin et al., 2007
Greater height	Martin et al., 2002
Improved eyesight	Chong et al., 2005
Lower risk of schizophrenia	Sorensen et al., 2005
Enhanced immune function overall	McDade, 2005
Benefits of Breastfeeding for the Mother	
Prevent postpartum hemorrhage	Heinig and Dewey, 1997 and references therein
Quicker return to pre-pregnancy weight	Heinig and Dewey, 1997 and references therein
Lower rates of breast cancer in women who breastfeed (Zheng et al., 2000)	Collaborative Group on Hormonal Factors in Breast Cancer, 2002; Heinig and Dewey, 1997
Reduced severity of anemia, bladder, and other infections	Labbok, 2001 and references therein
Lower risk of osteoporosis and associated fractures (Huo, Lauderdale, and Li, 2003)	Wolf, 2006 and references therein
Lower risk of atherosclerosis and heart attack	Stuebe, 2007
Lowered incidence of type 2 diabetes	Steube et al., 2006

have ear infections or to become overweight during childhood.[77] Despite increasing knowledge of the protective benefits of breastfeeding, thousands of women decide not to breastfeed or to terminate breastfeeding earlier than recommended, so this is not a trivial question to pursue. For many women throughout the world and for most of the past, the question of whether to breastfeed was not an issue whatsoever because there were very few other options available. Not only that, but breastfeeding mothers also populated their everyday environments, so there was likely not much apprehension about how to do it or whether the baby was getting sufficient milk. So the concern about promoting breastfeeding is somewhat recent, but it is no longer restricted to health-rich parts of the world.

As noted previously, the 2008 recommendations of the American Academy of Pediatrics are that babies should be fed exclusively breast milk in the first six months of life. More than 80% of new mothers in the United States initiate breastfeeding, but only about half of those continue to breastfeed for six months and almost all provide some form of supplementation long before that time. Furthermore, early supplementation usually leads to unhealthful feeding in subsequent months.[78] Why don't women follow the AAP recommendations? It is certainly possible that the message is not getting to them, and many women seek advice from their mothers and grandmothers who themselves may not have breastfed their children. Thus, unlike women in most of the world and in the past, there are few women who can serve as role models and even with support groups like La Leche League, day-to-day support for breastfeeding is limited.

The most common reason for supplementing or for stopping breastfeeding cited in study after study is insufficient milk or concern that the baby is not satisfied with milk alone. As I have shown, "insufficient milk" is usually a culturally induced phenomenon that is explained not by biology but by sociocultural context and behavior and their effects on biology.[79] For example, many women who were proud to see their infants fit and even exceed the growth charts for the first three months became alarmed when their 5- or 6-month-old breastfed infants began to fall behind in weight gain. As we have seen, breastfed babies do grow slower from 3 months on in comparison with their formula-fed counterparts who served as the reference points in older growth charts. Breastfeeding mothers who saw their babies fall to a lower percentile started to supplement or ceased breastfeeding altogether so that their babies could catch up with the charts.[80]

Other factors affecting contemporary breastfeeding behavior include the need to return to work,[81] social and physical discomfort with breastfeeding (for instance, sore nipples), and interference with normal lifestyle.[82] Entire books have been written on how to educate women about the benefits of and techniques for breastfeeding, so it is not necessary to delve into these now, but certainly it can be argued that difficulty with breastfeeding leading to early termination is a novel challenge that was almost unheard of in the evolutionary past, when such difficulties would likely have severely compromised the health of infants.

The list of reasons a woman may not breastfeed is endless. Near the top of the list is the lack of information and support for breastfeeding from doctors, family members, friends, and society at large. Having to return to work is often cited as a reason to terminate early, but it should not be that difficult to introduce supports in the workplace for breastfeeding mothers, especially considering the benefits to the workers, their children, and society at large.[83] Much attention has been paid to the idea common in the United States that breasts are sexual objects and displaying them in public for any reason borders on obscenity.[84] In fact, just as I was writing this section of this book, my local newspaper featured a front-page article about the social networking website Facebook pulling a photo of herself breastfeeding her infant that Kelli Roman had posted. The photo was apparently pulled because of a policy against "obscene, pornographic or sexually explicit" material. In protest against this interpretation of breastfeeding, Kelli started a web-based group that had grown to almost 100,000 members at the time of the article. Another group of protesters organized a "virtual nurse-in" with more than 11,000 people uploading photos depicting breastfeeding on Facebook websites.[85] Despite ongoing educational efforts, however, breastfeeding in public continues to be unacceptable in the United States. Certainly advertising, free samples, and questionable research from companies that manufacture milk substitutes also have to be listed near the top of any enumeration of reasons that women do not breastfeed.

People often say it is just too much trouble to breastfeed, which seems surprising considering the time and expense involved in preparing and using bottles to feed infants. Formula alone can cost between $700 and more than $3,000 for a year's supply,[86] depending on the brand chosen. The annual cost to provide formula to mothers enrolled in the U.S. government-funded nutrition program Women, Infants and Children (WIC) in 1997 was $2,665,715.[87] Despite efforts to promote breastfeeding among WIC recipients, these mothers have not breastfed at higher rates than non-WIC mothers in the past several decades.[88] Another estimate is that the approximate health care cost savings from breastfeeding is $4.18 billion.[89] A study comparing two groups of healthy infants in Arizona and Scotland found that among formula-fed infants there were more than two times as many office visits and days of hospitalization, and more than six times the number of prescriptions for lower respiratory tract illnesses, middle ear infections, and gastrointestinal illness compared with infants who were exclusively breastfed for at least three months. The estimated costs for these excesses were "between $331 and $475 per never-breastfed infant during the first year of life." [90]

Women often report that breastfeeding is painful and uncomfortable for them and that it actually interferes with their relationship with the infant because of the physical discomforts.[91] Almost uniformly, the 33 women interviewed in one study reported surprise at how painful and uncomfortable breastfeeding turned out to be, especially in the early weeks.[92] This issue is not commonly addressed in breastfeeding promotion literature, although several websites and pamphlets offer suggestions for making it easier and more comfortable for the mother. Engorgement, infections (mastitis), cracked and bleeding nipples, and uterine cramping were reported by several women in the study cited earlier, although these are rarely reported in other cultures where breastfeeding is more common. Some women who were interviewed found breastfeeding to be worse than labor and delivery, with regard to pain and discomfort. The women became scared of breastfeeding and rather than look forward to it as a quiet time for bonding with their infants they approached it with anxiety and trepidation. Not surprisingly, many women terminate breastfeeding when the discomfort becomes extreme. Unfortunately, many perceive that they are failures as mothers, and that, yet again, their bodies have not lived up to their or society's expectations. Feminist writer Jacqueline Wolf decries the inattention paid to breastfeeding by feminist scholars who have otherwise led the way for women's health reform in the last several decades.[93]

Why should breastfeeding be painful? In most cases the pain is due to the ways in which the infant is held and how he latches on,[94] factors that can be altered with information and guidance, especially from other nursing mothers and breastfeeding support organizations like La Leche League. Lotions, warm compresses, warm baths, and massage can help alleviate pain. In most cases, the pain diminishes with time if the mother continues to breastfeed and positions the baby correctly. Excessive breast tenderness and associated pain with breastfeeding may well be products of modernization and the overprotection of women's breasts from exposure. It seems unlikely that this would have been a problem in the past when women wore nothing across their breasts or covered them loosely.

Most decisions not to breastfeed or to terminate early revolve around psychosocial issues, but recently, a number of concerns have led to what might be called good medical reasons for not breastfeeding. The CDC and other health agencies in health-rich countries recommend against breastfeeding for women who are HIV positive because transmission of the virus from mother to child is known to occur. But in parts of the world where milk substitutes are unavailable or vastly inferior to mother's milk, the risks of dying from diarrhea and malnutrition may surpass the risks from HIV transmission, so advice against breastfeeding must take these other sources of morbidity and mortality into account before recommendations are made. In some places mothers *know* that their infants will die if they are not breastfed, whereas the possibility of the infant becoming HIV positive from breast milk seems more remote. Furthermore, there is evidence from a few South African studies that HIV-positive infants survive longer if they are breastfed.[95]

Some infants have enzyme "deficiencies" that may lead to intestinal problems if they are breastfed, although these are relatively rare. These include lactose intolerance, phenylketonuria (PKU), and a few other metabolic disorders that may require close monitoring but usually allow for partial breastfeeding. In the past, there would have been no other choice, most likely, so a few infants may have suffered lifelong consequences from subsisting only on mother's milk for the first few months. There are also rising concerns about environmental contaminants in milk, including those from herbicides (like Agent Orange), pesticides (like DDT), heavy metals (such as lead, mercury, arsenic), and radioactive agents (like strontium 90).[96] Most analyses suggest that the levels passed to infants through breastfeeding are lower than would be passed through milk from other species and water and, in any event, the benefits usually far outweigh the possible risks unless environmental contamination is widespread and extreme.

In this chapter I have made the case that breastfeeding has been an important component of mammalian reproductive strategy since the origin of the class several hundred million years ago. Most of this chapter has represented the views of those who advocate breastfeeding as providing the best food for young infants, but many who are reading this book were not breastfed and may not have breastfed their own infants; yet you may see little to no evidence that health has been compromised. This is further illustration of the flexibility of our species and how cultural mechanisms have allowed us to transcend some of the biological limitations placed on our ancestors and on many other species. Without breast milk, severely compromised health and even death were likely in the past. Maybe breast is best (a view I clearly hold), but it is far from the only option available to modern humans living complicated lives. I have presented evidence that breastfeeding plays a positive role in a woman's lifelong health as well as that of her infant, especially with regard to immune function. But a lot of women choose not to breastfeed or are unable to do so for very long. For many, rather than being able to provide the best for their infants, competing demands require that they can simply do the best they can and hope that it is "good enough." Of course women *are* more than their breasts and a mother's role in infancy is more than breastfeeding, as we will see in the next chapter.

8

But Women Are More Than Breasts

Young infant primates have sufficient motor skills to cling to their mothers soon after birth and can hold on while their mothers forage for food, run to keep up with the group, and sleep at night. The extra weight adds to a mother's daily food needs, but otherwise, an infant does not appear to interfere a great deal with her ability to carry on with her normal activities. As I noted previously, human infants do not have the degree of motor development necessary to support their own body weight, so they need to be supported by their mothers. Furthermore, their feet have lost the ability to cling because of adaptations to bipedalism, and their mothers do not have enough body hair for the babies to cling to even if they had strong muscles and prehensile toes. Because of the low nutrient density of their milk, the mothers do not have the option of leaving the infants in a nest or in the care of others unless those others are also lactating. They must carry the babies with them throughout the day, either in their arms or in a sling. If the mother had to use her arms to hold her infant, it would limit her ability to carry food or to use tools to acquire foods such as roots and tubers. On the occasions when she did have to put the baby down, she needed a way of keeping the infant quiet to avoid detection by predators. In the view of paleoanthropologist Dean Falk, the need for a mother to soothe the infant with her voice is the foundation of human language.[1]

The Costs of Carrying Infants

Given the costs and inconvenience of carrying infants, the "invention" of the infant sling would have been an important innovation in human evolution and one that allowed women to continue collecting foods even when they had young and helpless infants. In an experimental study, it was demonstrated that it is energetically more costly for a mother to carry her infant in her arms than in a sling.[2] Furthermore, the researchers were able to demonstrate that the costs of carrying infants for a female with a narrow pelvis, characteristic of the genus *Homo* in contrast to the genus *Australopithecus*, are even greater, placing more selective pressure on making tools for carrying infants as pelvic width narrowed in the course of human evolution.

When extrapolated over a year of foraging, they conclude that the costs of carrying an infant without a sling are greater than the costs of lactation.

The costs of carrying infants may have been one of the primary components maintaining birth intervals in ancestral humans. Anthropological studies confirm that women travel great distances carrying food even when they are encumbered with young children[3] (Figure 8-1). Consider that if a woman has an infant every two years, she would have to carry two infants over great distances, at considerable cost to her own energetic resources. Table 8.1 summarizes data that anthropologist Richard Lee collected in his studies of the !Kung, showing the average weight of children at

Table 8-1 Energetic costs of carrying children

Age of child	Average weight (kg)	Average kilometers carried per year	Kilograms/ kilometer
0-1	6	2,400	14,400
1-2	8.8	2,400	21,120
2-3	11.6	1,800	20,880
3-4	12.4	1,200	14,880
4+	13+	0	0

Modified from Lee, 1980.

Figure 8-1 !Kung San, or Ju/'hoansi, women carry home gathered foods, Kalahari Desert, Botswana. Marjorie Shostak/Anthro-Photo.

various ages, average distance that a woman walks in a year, and the cost of carrying infants of various ages for each of those kilometers. From this table it can be easily seen how costly it would be to have two children under age 4. And in fact, given the close association between energy availability and ovulation, it is unlikely that a forager woman could accumulate the reserves to ovulate until her infant approached age 3 and was able to walk part of the distance with his mother. For example, if an infant is born when his sibling is 2, the workload for the mother would increase to as much as 35,280 kilograms for a year of carrying the infant and the 2-year-old child. Lee argues that the two-year birth interval would be the theoretical limit to the ability of a woman to meet her own and her infants' energetic needs. At this rate, she may not be able to support both children, suggesting another point where trade-offs result in longer birth intervals if they are associated with healthier offspring. Of course, if she can get help from other group members (her mate, older children, or older, non-reproductive females), she would be able to maintain a lower birth interval.

Human infants need to be carried, but at some point, children begin moving about on their own. Locomotor skills are among the "developmental milestones" that parents and health care providers pay attention to in the first year or two of an infant's life, and crawling as an expected milestone for a child between 7 and 12 months of age appears on almost all lists. In fact, it is regarded as a universal characteristic of infancy, and parents often become concerned if their child is not crawling by his first birthday. Crawling is seen as an important stage toward full bipedal walking and one necessary for the development of strength and motor control. But anthropologist David Tracer argues that infant crawling is a somewhat recent phenomenon and may well be associated with the appearance of the first floors in houses and hygienic practices that reduced exposure to parasites and predators.[4] Consider that an infant who is placed on his belly on the ground is exposed to a lot of noxious agents including human wastes, parasites, hot fires, and various pathogens that can cause diarrhea and other, potentially worse, diseases. What Tracer found to be more common in many indigenous cultures, including the Au people that he studied in Papua New Guinea, was carrying for a longer period of time and placing the child in an upright, seated position when he was awake but not being carried. The phase that precedes walking in this population is upright "scooting." In fact, crawling is actively discouraged by Au parents. As a consequence, Au infants "fail" the horizontal tests in the widely used Bayley Scales of Motor Development, but do quite well in the vertical positions, the opposite of what is seen (and judged as "normal") in Western populations. Yet again we have evidence that what is normal in one culture may be quite the opposite in another.

Sleeping with Infants

As noted, constant contact includes sleeping with the infant and breastfeeding throughout the night, certainly a component of reproductive strategy for humans and most other primates. It is part of the human evolutionary legacy that infants "expect" to be with their mothers at all times. Consider what would have happened to an infant in the evolutionary past who was put down in another part of the sleeping area while the parents tried to ignore the cries throughout the night. It makes sense from every possible angle to argue that mothers and infants slept together throughout history, just as they do in most traditional cultures today. But the American Consumer Product Safety Commission, well-meaning pediatricians, and the general public in the United States have different views. Many health care professionals and parents consider that sleeping with your baby can be dangerous and can lead to a myriad of disorders from death (due to sudden infant death syndrome/SIDS or parents rolling over and crushing the baby) to poor psychosexual development. Anthropologists Jim McKenna and Thom McDade suggest that rather than being a source of nighttime comfort and sustenance, the mother's body is often seen as a "lethal weapon."[5] An evolutionary perspective presents a different view.

Jim McKenna is probably the best-known advocate of what he refers to as the "biologically appropriate sleeping arrangement" for human mothers and infants.[6] In fact, it should be obvious that nighttime breastfeeding is nearly impossible if the infant and mother are not in close proximity. As noted earlier, ancestral infants who were put aside during the night were very likely not to survive until morning due to cold, predation, and hunger. So if breastfeeding throughout the day and night is part of the human legacy, then so is co-sleeping. Furthermore, despite all sorts of negative press, mothers in the United States choose to sleep with or very near their infants in huge numbers, especially if they are also breastfeeding. When they do sleep with their infants, mothers report that the infants cry less, both sleep better, and she produces more milk.[7] McKenna's research shows that mothers are more responsive to infant arousals and would be more likely to detect breathing problems that the infant might experience, thus *preventing* rather than *causing* SIDS.[8] In his view, babies are more likely to die from not sleeping with their parents than when sleeping with them, except in cases where a parent is under the influence of alcohol or drugs. He believes that in most instances, SIDS is a culturally influenced cause of death that is rarely associated with biology. Although McKenna's research does not allow an actual recommendation of bedsharing, it does reject universal recommendations against the practice.

As discussed earlier, one of the most frequent complaints of new parents is inability to get sufficient sleep when a baby joins the household. McKenna notes that the phrase "good baby" is "practically synonymous with a baby's ability to 'sleep through the night' alone."[9] In fact, a question about sleep is one that is frequently asked of a new mother by pediatricians and friends in the United States, indicating an almost universal expectation that sleeping through the night alone is a milestone to be achieved as soon as possible. It is also clear that early independence is a behavior encouraged by Western societies like that of the United States, and solitary sleep serves to implement that goal. McKenna notes that the solitary-sleeping infant who quickly "learns" to sleep through the night is both the valued and the "normal" infant, although it is a far cry from human infants of millions of years of evolutionary history.

The evolutionary perspective has successfully argued that feeding human infants on a four-hour schedule not only does not make sense but is unrelated to mother and infant biology. Gradually this perspective is also changing the way parents and pediatricians view infant sleep. Obviously, there is a simple and time-honored solution to the "problem" of infants not sleeping through the night.

Other aspects of infant sleep in addition to location can benefit from examination through an evolutionary lens. For example, there is great concern about the dangers that result from infants being placed in bed on their stomachs. In fact, the Back to Sleep campaign was initiated in 1992 by a coalition of medical organizations in an effort to quickly educate caretakers about how to place their infants in beds, based on findings that infants who were placed on their stomachs were at higher risk of dying from SIDS. It has been quite successful, resulting in a 50% reduction in SIDS deaths since it began.[10] But as McKenna points out, breastfeeding mothers do not need slogans to tell them how to place their babies in bed; babies placed on their stomachs have great difficulty getting in a position to nurse, so a mother who wants to facilitate nursing is much more likely to place her infant on his back. Mothers who are breastfeeding seem to instinctively sleep on their sides with their legs drawn up in a way that partially encompasses the infant and would prevent her from rolling over on her baby.[11] When babies sleep in close proximity to the mother, they exhibit more arousals, more frequent breastfeeding, increased heart rate and body temperature, more movement and awakenings, and less time in deep sleep.[12] In fact, the deep sleep that most of us regard as desirable is far from the ideal for infants in the first few months of life when sleep interruptions (*apneas*) are frequent and failure to awake easily in response to them can result in suffocation and death.

It has been frequently noted that there is no animal model for SIDS. In other words, this seems to be a disorder unique to humans, perhaps to modern, Western humans who do not sleep with their infants. But the fact that it can occur in human infants and apparently does not occur in other species suggests an underlying anatomical or physiological predisposition. Among the places to look for these are the developing brain and the respiratory anatomy. Between two and five months of age, the time when SIDS occurrence is most likely, self-regulatory behaviors (like breathing control) are changing from being reflexively to voluntarily controlled. The infant before this time apparently lacks "natural" responses to things that block breathing (such as mucus or bed covers) and must depend on environmental experiences to "learn how to breathe" and maintain breathing passageways. What better environmental cues are there than the regular, rhythmic breathing of the mother and the carbon dioxide that she expels close to the baby's face? When the infant sleeps alone, those tactile and auditory cues are absent, and breathing irregularities may become life threatening.[13] The fact that SIDS is more common (and may be unique) in humans than in other species is likely due to two factors: (1) the neurologically underdeveloped human brain; and (2) cultural factors that lead to infants sleeping alone.

Another biological factor influencing SIDS may be the structure of the upper respiratory tract, which is different in humans compared to other mammals. An important difference is that in humans, the breathing (larynx) and the swallowing (pharynx) channels converge rather than remain separate.[14] This means that humans cannot breathe and swallow simultaneously, as other animals can. But because this inability to breathe and swallow at the same time would prevent nursing from occurring, the configuration does not appear until an infant is about 2 years old, about the time that breastfeeding frequency decreases. In other words, the human infant looks like all other mammals with regard to the configuration of the breathing and swallowing passageways. Initial changes in the configuration begin between 4 and 6 months of age, the time when SIDS rates are highest. Anthropologist Jeffrey Laitman noted that this "may be a time of potential respiratory instability due to a changeover from one respiratory pattern to another."[15] The unique configuration appears to be related to bipedalism and upright posture, which explains why it does not appear until the human infant is a biped. Furthermore, although it seems to cause numerous medical problems (many people die of choking every year), the enlarged region where the two tubes converge allows for the almost-infinite array of sounds associated with human speech.[16] No matter how smart your infant seems to be, he or she will not have full language abilities until that convergence occurs.

One of the arguments used by pediatricians for urging that infants sleep alone is that only this will lead to healthy sleeping behavior later during childhood, adolescence, and adulthood. But not being able to get a good night's sleep is cited as a problem for 62% of contemporary American adults who were presumably "taught" good sleeping behavior by being placed alone in their cribs.[5] I wonder if the statistics would be different had parents and infants slept together for the last several decades rather than apart. It seems that if sleeping alone as infants were the "solution" we would not see so much evidence of sleeping problems in people of all ages. In fact, it is much more common for parents in the West to report sleep problems in their children than those in cultures where co-sleeping at all ages is common. Finally, by almost any measure (self-esteem, sociability, life satisfaction), co-sleeping has been found to have more positive outcomes than solitary sleeping in studies that have been conducted in the past two decades.[17] As McKenna notes, this is evidence that the benefits of co-sleeping may be as great for 21st-century people as they were for our ancestors. Of course, if the beds in which we 21st-century folks sleep have huge fluffy comforters or duvets, lots of pillows, are waterbeds, or include people who are obese, smoke, or are under the influence of alcohol or drugs, it may not be a good place to sleep now or in the past, no matter the other benefits.

Weaning and Beyond

Weaning is another topic that has been examined through the lens of evolutionary medicine. At some point, the reproductive "goals" of the mother and of her nursing infant come into conflict again when it is in her best interest to wean so that she can reproduce again—but it is in his best interests to continue deriving nutrients and immune factors from her for as long as possible. That point varies across species and across human populations, and there is variation in what may be regarded as the "normal" period of breastfeeding. A lot depends on the age of the mother (and thus her reproductive potential), the amount of pathogens in the environment, and the availability of appropriate foods for newly weaned infants. Anthropologist Dan Sellen proposes that weaning can come younger in humans than in comparative primate species because we have the ability to prepare what he calls "transitional foods" for feeding infants as they are being weaned.[18]

Among chimpanzees and most human foraging cultures, three to four years seems to be the typical period of infant nursing. Anthropologist Katherine Dettwyler[19] has examined a number of life history traits to determine what she calls the "natural age of weaning" for humans. For example, larger animals tend to nurse their infants longer relative to gestation length (for gorillas and chimpanzees the ratio of nursing to gestation length is 6:1) which would suggest that the weaning age for humans is 4.5 years. In many species of monkeys and apes, weaning occurs at the time of eruption of the first molars, which would be about 6 years in humans.[20] Dettwyler reports that primates wean their young when they have reached about a third of adult body weight, about five to seven years for humans. For those of us in cultures where breastfeeding, if it occurs at all, often lasts less than one year, these numbers (four to seven years) seem excessive, but for people in parts of the world where access to appropriate and healthful infant foods is limited, nursing for several years (with supplementation from other sources after six months) may mean the difference between a healthy and a sickly childhood.

On the other hand, anthropologist Gail Kennedy argues that the developing human infant brain cannot be supported on mother's milk alone beyond one year,[21] so we should expect supplementation and even weaning by that time if there are sufficient high-density foods available. Great apes can "afford" to breastfeed their infants for a long time, thus protecting them from malnutrition and infection, because they do not have to support the growth of a hungry, metabolically expensive brain. Humans, however, have a dilemma, according to Kennedy, in that they must sacrifice the immune protection of mother's milk for higher energy foods to support brain growth. By about 3 years of age, the brain simply cannot continue to develop with mother's milk alone or even with some supplementation, so safety is sacrificed with a transition to adult foods. If this strategy is the result of natural selection, it suggests that survival alone is not sufficient to ensure human reproductive success; rather, intellectual development and accompanying social and technological skills have equal impact. Of course, 3-year-old infants are not able to secure high-quality adult foods for brain development without extensive assistance and direct provisioning from parents and kin. Kennedy argues that the shift from the ape 5-years-of-nursing pattern to the human 2.5-years-of-nursing pattern occurred with a "tool-assisted dietary shift" about 2.5 million years ago,[22] associated with an increase in consumption of foods of animal origin. Help from kin, especially maternal grandmothers, may have been crucial to the success of this transition.[23]

No matter when it occurs, weaning is far from the end of parental care. As noted, one characteristic that distinguishes humans from other mammals and most other primates is that we continue to provide food for our young following weaning.[24] For most young mammals, once they are weaned, getting food is pretty much up to individual effort. Sharing of food, even between mother and child, is rare in other species. Anthropologists Chet Lancaster and Jane Lancaster suggest that the behavior of provisioning children between weaning and puberty may have doubled or even tripled the number of offspring that survived to adulthood for early humans. This long period of extended child care by older children and adults probably enhanced the time for learning technological and social skills, also contributing to greater survival and reproductive success. It also provided time for continued brain development, which is mostly finished by age 6.[25] Thus, the costs of extensive parental care may have been outweighed in human evolutionary history by the benefits of greater reproductive success for mothers and for the offspring who were provisioned.

Furthermore, Dan Sellen has suggested that the practice of preparing foods for infants as they are being weaned served as both a protection for the infants and as a way of reducing energy costs for the mother, leading to increased reproductive success for both and adding to the behavioral flexibility of human adaptation.[26] He argues that this may account for the lower birth intervals seen in humans in comparison with our closest primate relatives. But if the "complementary foods" are inappropriate, as they often are today, the strategy may backfire, resulting in poor infant and child health.[27] Sellen argues that interventions that target increasing infant and maternal health need to take into account the behavioral flexibility of infant weaning and the tendency to wean too early even when complementary foods are not adequate.

Future Motherhood and Breasts

Getting information out to the general public (especially those who are pregnant or planning to get pregnant) about the benefits of mother's milk for babies has apparently been so successful that human inventiveness and modern technology have come up with ways that moms can provide their milk to their infants when they are not able to actually breastfeed them. In this way, they can avoid being tied down to a breastfeeding regimen that is incompatible with other aspects of their lives such as working outside the home. Jill Lepore writes in the New Yorker that mechanical breast pumps are almost as ubiquitous as cell phones for mothers of young infants[28] and are even advertised as appropriate gifts for baby showers. For women who must return to work soon after their babies are born, these pumps are welcomed because they enable them to continue to provide what they regard as the best food they can offer their infants. Perhaps "breast is best," but mother's milk without the breast is at least better than most alternatives when a woman has no other choice.

This is not to say that pumping is easy or convenient, and in fact most women who try to pump while they are at work end up abandoning the effort because it is complicated and so little support is available. In certain jobs (usually low-paying ones where workers have little autonomy) pumping may be impossible because there is no place to store milk safely and no place to pump with privacy. Most women who work outside the home and choose to breastfeed their infants would prefer that their babies were with them at work, at least when they are less than 1 year old. Ideally the pro-family values touted in the United States in recent years would include pro-family work conditions that enable, rather than interfere with, breastfeeding. Lepore's concern is that by promoting breast pumps and even allowing time for pumping at work, we are "avoiding harder—and divisive and more stubborn—social and economic issues" [29] about parenting and the needs of women and their families.

As we have seen, breastfeeding is far more than a food delivery system, no matter how beneficial the food is for infant development. Psychologist Harry Harlow demonstrated decades ago that infants want soft, warm mothers in addition to milk. The sad little rhesus macaque babies he studied spent almost all of their time on the milk-less cloth "mothers," leaving them only to quickly nurse from the wire "mothers." But that having been said, and despite much research that suggests health throughout life is positively impacted by breastfeeding, public health campaigns that focus on women and blame them for not breastfeeding miss the point that decisions about infant care are often out of individual control. [30] I have written a lot about trade-offs in this book. Is a teenage girl who opts to return to school and leave her infant in the care of her mother who feeds her formula making a "bad" decision? She trades the good qualities of her milk for the chance to improve her social and economic options through education. Which choice is most likely to improve the long-term health of her child? In the United States, education probably trumps mother's milk in this case. In a pathogen-infested impoverished home, staying home and breastfeeding may be the better choice for infant survival.

How about a woman who is the sole breadwinner for her family? If she does not return to work soon after birth, she and her family may lose their home, resort to subsisting on foods of inferior quality and quantity, and be unable to afford needed health care. Quite a trade-off for even the best things that mother's milk does for maternal and infant health. Then consider the severely depressed mother who may be able to provide milk but does so under duress and for whom breastfeeding worsens depression or increases stress. I have presented evidence and arguments based on evolution that, all other things being equal, breastfeeding is the very best option for feeding infants, but humans throughout evolution have rarely been able to pursue the optimal or best strategy. Far more common is the "good enough" strategy. Under circumstances where breastfeeding interferes with family health and well-being, formula feeding is certainly good enough, especially if the warmth and emotional interaction of breastfeeding is provided by the person holding the bottle. If breastfeeding is problematic because it interferes with golfing, partying, and horseback riding, maybe it would be advisable to reconsider the options.

Another technological fix that seems to derive from concerns about parenting is a machine that monitors an infant's cry, interprets it, and informs the parents what the infant is trying to say with his cries. In other words, it translates baby communication into a "language" that parents can understand; it solves the problem that arises from a crying baby who "is no longer merely a being to be loved, but a problem to be solved."[31] For about $100, you can buy a calculator-sized device that will interpret cries and categorize them into five possible meanings: hungry, sleepy, uncomfortable, stressed, or bored. To work correctly, the device needs to be positioned about two yards from the baby (the device comes with a chart that tells you how far to place it depending on the baby's weight) and it takes about 20 seconds to produce a translation.[26] A light on the monitor tells you that the signal is being processed and changes color when the diagnosis is complete. If you are not sure of the diagnosis, you can look at the baby's body language, consult another chart, and come to a conclusion. Presumably, once the parents know why the baby is crying, they can respond appropriately (there is yet another chart that lists possible ways of calming the baby), although the delayed response may result in a different type of cry (a change from bored to stressed). Many agree that this gadget has nothing to do with communication, which is a two-way process between infant and parent[32] and which is at the base of what it means to be human. For those who say that the device will reduce parental anxiety, I offer the suggestion that a few days spent with the baby, communicating with him in a two-way fashion, would probably be a better way to overcome most anxieties about what the baby wants when he cries.

Another example of where technological innovation seems misguided is the one-way baby monitor that is placed in the infant's room while he or she is sleeping, enabling parents to monitor the sounds the babies make while they cook, clean house, read, sleep (in their own rooms), or otherwise carry on their daily lives. Jim McKenna claims that based on our understandings of the evolutionary and cross-cultural environments of infancy, the monitor, if it is used at all, should be turned the other way.[33] He argues that infant sleep, arousals, heart rate, and breathing all benefit from hearing the sounds that are part of normal living. Of course, the argument from an evolutionary perspective as discussed in previous chapters (and extensively by McKenna in many of his publications) is that infants should be physically with their parents almost all of the time, in which case exposure to the sounds of everyday living is a given. But the reality of many of our lives is that we cannot keep our babies with us at all times, and we may not feel comfortable having the baby in our beds or even in our rooms at night. In this case, a monitor that broadcasts both ways may be the solution. Parents can hear the baby and respond if necessary and he can hear the sounds his parents and other family members make.

Perhaps not so egregious, but still a long way from ancestral infancy, are products aimed at increasing an infant's intelligence by exposing him to music, "educational" videos, television, and other passive stimuli. Some of the advertising for these products claims that it will increase intelligence, but most of the research suggests otherwise. In fact, not only is there little evidence of positive effects on intelligence, but there is more evidence of potential harm to infant development from too much exposure to visual media.[34] Early language development and other aspects of intelligence proceed well enough in the context of visual, aural, and other aspects of communication with a real person, especially parents.[35] This was the context in which infant skills developed for several million years of evolutionary history.

In evolutionary medicine, a question that is often asked is, "what is the evolutionary environment" of a given behavior—meaning what were the ancestral conditions under which it may have evolved? For infants, the evolutionary environment was and is the mother's body.[36] Any measure of infant development that hopes to reveal normalcy must consider how it occurs when the infant is in contact with the mother, being held or carried by her, breastfed by her, or sleeping with her. Under these conditions, infants feed frequently, rarely cry, almost never die from SIDS, sleep on their backs or sides so as to facilitate nursing, awake frequently and fleetingly in the night, and grow up to be healthy children and adults.

Because human infants are relatively helpless at birth and dependent on parental care for as much as a decade or more, a mother's commitment of time and energy does not stop when she weans her child, although after that, other family and community members can provide comparable care. In fact, caretaking assistance from others was likely an important key to the mother's ability to bear another child and to begin the intense direct-investment-in-infants cycle again. Evidence from studies of other primates and human foragers indicates that infants are in near-constant contact with their mothers or another person for the first three to four years of their lives. Consider that a mother who carries her infant while she forages must obtain enough calories to support her own metabolism and energy expenditure, including that required to carry her infant, and to produce sufficient milk. As the infant grows, he or she gets heavier and heavier until the time arrives for weaning, after which time others can assist in child care. Once the baby is weaned and the mother is freed from lactation and carrying an infant, if she continues to get approximately the same amount of calories, sufficient reserves can be built up so that ovulation occurs again, leading to pregnancy, breastfeeding, and carrying a new infant. Support from others (especially grandmothers, as we will see in the next two chapters) and shared child care are key to this strategy, and probably were at the base of the evolution of the family structure and extended kin network that characterizes humans and distinguishes us from most other mammalian species.

9

If Reproduction Is What It's All About, Why Does It Stop?

An evolutionary perspective on women's biology across the life span ultimately runs up against menopause, a phenomenon that does not initially make sense if the currency of evolution is reproductive success. If increasing reproductive success is what evolution is about, then how could natural selection favor terminating reproductive functioning relatively early in a woman's lifetime? In the historic and prehistoric past, women may not have lived very many years beyond menopause, but today this event occurs when women have as much as one third of their active and healthy lives ahead of them. Why have we not seen an increase in the age of menopause just as we have seen a decrease in the age of menarche as health conditions have improved in the last 200 years? In fact, there are two questions to address when considering menopause from an evolutionary perspective: why does ovulation cease at all, and why do women live so long after they stop reproducing? I will address the first in this chapter and save postmenopausal longevity for the next.

Is Menopause Unique to Humans?

In the medical literature, menopause is the end of menstruation, defined retrospectively after a full year has passed without a menstrual cycle. As noted in Chapter 2, however, this definition is based on the assumption that the "normal" state for a woman is menstrual cycling. As we have seen, for most of human history women were pregnant or breastfeeding children and were rarely menstruating, so paying attention to how long it had been since the last menstrual period may not have been meaningful for explaining any biological processes. Because most other mammals do not menstruate (see Chapter 2), it is meaningless to ask if they have a menopause. Thus, it is safe to say that menopause, when the definition depends on regular menstrual cycling, is unique to humans. As has been pointed out, however, the only meaningful definition of reproductive termination that works across species is the cessation of ovulation[1] and this is the more common meaning found in the study of primatology.

To understand how other species experience the end of reproductive cycling, it is perhaps more useful to ask if there are other examples of females who regularly cease ovulating before death. Virtually all female mammals that live long enough show signs that look like the approach to menopause for humans: increasing length of ovarian cycles and percentage that are anovulatory; increased interbirth intervals; decreased sexual activity; and biochemical measures like hormonal increases and decreases, egg depletion, and decreased bone mass. It is argued that by these measures, and looking at menopause as a process rather than an event, chimpanzees and many other primates serve as good models for human reproductive decline.[2]

Another line of inquiry is to examine whether other species have long postreproductive lives. Although some degree of postreproductive life has been reported for many mammals (including some species of whales and the Asian elephant),[3] no other primates live such a high percentage of their lives after reproductive termination. Female chimpanzees and monkeys experience decreased fertility in their later years, but most continue cycling until their deaths.[4] Furthermore, in cases when they do cease ovulating before death, they are usually in very poor overall health, indicative of being much further along in the aging cycle than human females usually are when they cease ovulating.[5] Primatologist Toshisada Nishida reports that at least five female chimpanzees at Mahale survived an average of 10 years after they last gave birth,[6] but despite occasional reports of extended life beyond reproduction in apes and monkeys, it is far from the routine and expected phenomenon that it is in humans.[7] One estimate is that fewer than 5% of chimpanzees and baboons in the wild live past the age of reproduction; even in human foraging populations, as many as one third of adult females live beyond age 50.[8] Lee estimated that 10% of the !Kung that he worked with were over 60 and most of them continued to make significant economic contributions to their bands.[9]

Why Cease Reproducing at about Age 50?

Perhaps we cannot defend menopause as a unique feature of humans, but that still leaves open the question of why it exists at all in any species, given the value placed on increased reproductive success. One theory about menopause is that it is just the result of the extension of the human life span, a by-product that has been "uncovered" in the last several hundred years as longevity has increased. This view claims that the maximum life span of the mammalian egg is about 50 years (remember, all the ova are already present before birth) and that no mammals can reproduce after the eggs have been depleted, no matter how long they live.[10] By this reasoning, although the human life span has increased over the past several hundred years, the reproductive life span has not because the eggs simply cannot live much longer and the ovary just runs out of eggs. In the language of evolution, cessation of reproduction at about age 50 is an ancestral characteristic that we share with most other long-lived mammals, but a long postreproductive life (for women) is the derived or recently evolved characteristic that appeared after humans and apes diverged from a common ancestor.[11] With regard to menopause, we are like apes. With our long lives, we are an unusual primate.

Another proposal to explain menopause is that so much energy is needed "up front," for reproduction in the early years, that there is nothing left over by the time a woman reaches 50. This is the concept of pleiotropy, which argues that genes that increase survival and reproductive success at early ages will be favorably selected even if they have negative effects on survival at later ages.[12] Anthropologist Jocelyn Peccei proposes that each new infant is increasingly costly to a mother, especially for humans who produce large-brained and extremely dependent infants who need direct maternal care via lactation for three to four years. Even a few infants will deplete a woman's energetic reserves to the point that her reproductive success may be greater if she invests her increasingly limited resources in her most recent offspring rather than giving birth again.[13] Because it takes about 12 to 15 years for a child to become independent, it is argued that females who lived that long had greater reproductive success than those who died before their children entered their teens.[14]

Early theories for menopause proposed that it was the object of selection itself because women who ceased reproducing early would have greater reproductive success through providing higher quality care for their current young than they would through having more, potentially lower quality young, who are less likely to survive. This has been difficult, however, to explain mathematically, based on the assumption that producing one or more offspring that share 50% of your genes would always be better than helping other young, even if they had as much as 25% of your genes.[15] Furthermore, anthropological evidence is inconsistent with the idea that older women provision juveniles at levels that would increase fitness more than having their own offspring.[16] Finally, if ceasing to reproduce early is advantageous if it leaves you with 15 to 20 years of healthy life left, it suggests that the age of menopause would have gradually increased in the past 100 years as living conditions and health have also increased. Despite incontrovertible evidence that these improved conditions have driven the age of onset of reproduction downward in almost all populations, there is no evidence that the age of termination of reproduction has been increased in the same way. In order for natural selection to act, there must be variability in a trait. Certainly there is variation in the age of menopause, but the average age is somewhat stable across human populations and does not show evidence of trending upward, even in health-rich populations. This suggests that cessation of reproduction is something that we, as mammals, are "stuck with," and that it is part of our heritage that has limited ability to respond to natural selection afforded by improved ecological circumstances.

As noted in Chapter 5, childbirth is challenging for bipedal humans who give birth to large-brained infants, and the risks are not negligible.[17] It is estimated that more than half a million women die of complications related to childbirth every year, and the risks increase at older ages with women over 40 having five times the risk of dying in childbirth as women 20 to 24.[18] By another estimate, at ages greater than 40, the risk exceeds 30 times that at ages 20 to 29.[19] Given these risks, it makes sense that pregnancy and birth would become increasingly rare beyond age 45, to the point that the chances of dying in childbirth exceed the chances of giving birth to a healthy infant, say at about age 50. Furthermore, a 50-year-old woman several hundred years ago would have probably been in poorer health than her equivalent today (at least in health-rich populations), suggesting that the risks may have been even greater. In particular, the risks of postpartum hemorrhage and blood clots increase with age, and placental and uterine problems increase with number of pregnancies. If a 50-year-old woman gives birth, she and her infant may die. If she dies, chances are that her other young children will also face compromised health and may even die or fail to reproduce. Thus, one could argue for cessation of ovulation early enough in life to ensure that you had several good years left for raising your young. Couple this with an increased ability to care for your grand-children once freed from reproducing yourself, and the selective value of ceasing to reproduce may outweigh the negatives at the population or species level.[20] There is far from agreement on this proposition.[21]

As I have repeatedly emphasized throughout this book, maximum fertility (quantity) is not necessarily the best way of achieving reproductive success. Humans, with such huge demands placed on them for raising offspring, seem to have adopted a strategy that puts quality before quantity, meaning that the optimum number of offspring is far below what can theoretically be achieved.[22] This strategy appears to be particularly successful in situations where resources are constrained and there is a lot of competition for them, as probably characterized most of human history and most populations today. The same argument can be applied to reproducing beyond age 50.[23] As already noted, energy limitations make it selectively advantageous for women to stop cycling when conditions for preg-nancy are not good. These include athletic amenorrhea, illness, psychosocial and emotional stress, starvation, and, perhaps, being older than 50 years.

Factors Affecting the Onset and Experience of Menopause

Menopause is directly determined by the number of eggs a woman is born with, but several factors seem to influence its timing, perhaps by influencing the rate at which the eggs decrease through time. Geographic location, income, education, and marital status have been related to age of menopause.[24] For example, single women experience earlier menopause than married women.[25] This effect may be related to sexual behavior, social environment, or other psychosocial aspects of the marital bond such as social support or stress reduction. One group of researchers suggests that the mechanism may be pheromonal—the presence of a man and male pheromones may influence ovarian function and estrogen levels in ways that delay menopause.[26]

Factors that seem to be related to earlier age at menopause include childhood undernutrition, low socioeconomic status, low education levels, rural living, never having given birth, short menstrual cycles, and lifestyle factors such as smoking and alcohol consumption.[27] All of these factors are interrelated, of course, so it is not possible to say which has the strongest effect or if any have independent effects. Age of menarche and use of hormonal forms of birth control do not seem to impact age of menopause. How parity influences age at menopause is unclear, with some studies showing an effect and others showing none. High dietary intake of fat, cholesterol, and coffee were associated with later menopause in a Japanese population.[28]

A number of lifestyle and environmental factors seem to have effects on symptom reporting at menopause and include body weight, body constitution, and exercise.[29] A number of studies have found a positive association between BMI and risk for moderate to severe hot flashes compared with low BMI.[30] Aerobic exercise can reduce both psychological and physiological responses to psychological stress.[31] There is evidence that a woman's experiences with and attitudes toward sexuality may affect her experience of menopause.[32] There is also a reported association among increased irregularity of menstrual cycles, hot flashes, declining estrogens, and declining frequency of sexual intercourse.[33] There is a report of associations between tobacco usage and both severity and frequency of hot flashes.[34] Current smokers were more likely to experience hot flashes than past or never smokers in another study.[35]

In some cultures it is believed that menopause is an emotionally and psychologically challenging time for women, and several studies have examined the relationship of mood and other psychological variables to the menopause transition.[36] Depression, for example, has often been reported by women during this time, but most studies suggest that hormonal changes are not necessarily responsible.[37] In one longitudinal study[38] low mood and physical symptoms were shown to co-occur in midlife but have different causes. In two other studies, earlier life experiences were better predictors of distress at midlife than menopausal status[39] and women approaching menopause were more susceptible to developing depression than were younger women.[40] Socioeconomic status was associated with symptom reporting in a large multi-city, multi-ethnic study of 16,065 women in the United States. Those who reported difficulty paying for basic needs also reported more symptoms.[41]

Finally, there are many differences in the experiences of women during the menopause transition across cultures, even in the "physical symptoms."[42] Hot flashes, for example, are reported by many women in North America, Thailand,[43] Norway,[44] India,[45] Nigeria,[46] and Tanzania[47] but are rarely reported for Japanese women,[48] Navajo women,[49] Mayan peasant women,[50] and Sikh women living in Canada.[51] This is further evidence that the relationship between biological aspects of the menopause and psychosocial and physical aspects is mediated by the lived experience of each woman, not the least of which is her cultural milieu.

Endocrinology of Menopause and the Late Reproductive Years

The reproductive life spans of women and men are quite different: in women, reproduction stops somewhat predictably at about age 45–50. Although the ability in men to reproduce declines with aging, it does not completely cease, at least at a predictable age. Evolutionary biologist George Williams suggested that this difference is due to the relative costs of reproduction to the sexes, with the costs to the female generally being much greater than the costs to the male. Moreover, the risks for an older woman from pregnancy and birth are greater than the risks to males producing sperm, and as we have seen, they may be greater than the benefits.[52]

Reproductive hormones like estrogen and progesterone production begin to decline toward the end of the reproductive years until ovulation (and thus menstruation) ceases altogether. Recall that after ovulation, the corpus luteum secretes estrogen and progesterone, which inhibit FSH and LH. At the end of the reproductive years, there is no egg, and so there is no corpus luteum, nor high levels of estrogen and progesterone; and LH and FSH are not inhibited. This means that FSH and LH gradually rise and remain high in the postmenopausal years and it is these hormones that can be used to signal that the transition from reproductive to postreproductive state is underway.

In previous chapters I noted that highly frequent menstrual cycling is implicated in various reproductive cancers, so it seems that the cessation of cycling with menopause may be protective because it puts an end to high exposure of breast tissue to estrogens and high cell turnover rates over the course of 30 years of menstrual cycling with few interruptions.[53] If 30 years of cycling put a woman at risk of breast cancer, imagine what 40 or 50 years would do. It is also important to consider the energy costs of continuing to cycle late in life.[54] As noted earlier, each successive pregnancy uses a greater proportion of maternal energy reserves, meaning that the sixth or seventh pregnancy is relatively more expensive than the first or second.

Is Menopause a Medical Concern?

It appears that menopause is simply the end of reproduction for women, a phenomenon that potentially occurs for all mammals and is not necessarily something that needs to be "explained." But why does menopause have to be a "bad" thing? This question derives from a fairly Western perspective that has pathologized the period of time when a woman is experiencing the physical and emotional transitions to a postreproductive state.[55] In fact, the idea that menopause is an illness that needs to be treated has been part of Western understanding for more than 200 years.[56] It is often likened to diabetes as a disorder characterized by hormonal deficiency. In fact, menopause has been called an "estrogen deficiency disease."[57] Furthermore, it is often listed as a risk factor for a variety of diseases and disorders of aging,[58] reinforcing the idea that it is a medical "problem."

Much of the medicalization-of-menopause phenomenon is based on the idea that women did not live longer than about age 50 in the past and that, therefore, prolonged life is "abnormal" and requires medical intervention to maintain health.[59] This was part of the rationale for having women take hormone replacement therapy to enable their reproductive hormones to last as long as their lives. According to this view of menopause as an "estrogen-deficiency disorder," external sources of estrogen needed to be added back into the system to keep mental and physical health at levels seen during the reproductive years. If you provide insulin to diabetics so they can maintain healthy lives, why wouldn't you provide estrogen to postmenopausal women? An obvious answer is that diabetes is not a universal phenomenon of the human life course, whereas cessation of ovulation is. This life course approach to menopause leads one away from seeing it as a medical problem resulting from a "deficiency."

But recasting menopause as a developmental rather than a degenerative phase does not help women who experience stressful symptoms they perceive to be associated with menopause. Horrible hot flashes that keep you awake at night and interfere with your daily activities are not going to be relieved simply by positive thinking or mantras that "this is normal, this is normal." Like everything else that is evaluated from an evolutionary perspective, a normal, healthful physiological response (such as elevated body temperature and withdrawal of iron when exposed to disease) can become abnormal and even pathological when it is excessive. But a modification in the way of thinking about the problems many women experience at menopause may help to separate aging from normally occurring hormonal changes. Medical interventions may be necessary and desirable to treat osteoporosis, hot flashes, and cardiovascular disease, but they are treating these specific disorders, not menopause per se.[60] In fact, useful and appropriate treatments are much more likely to be developed when the specific disorders are considered rather than focusing on the life phase transition that all women experience if they live long enough. In this way, the problems will be treated, not the life phase. This will avoid the problems that occurred when all postmenopausal women were advised to adopt HRT.

Often terms used to describe menopause and the menopause transition reinforce the idea that it is a medical issue, including using the term *symptoms* to describe phenomena such as irregular menstrual periods and vasomotor changes. As noted in previous chapters, anthropologist Emily Martin and others have pointed out that the metaphors we used to describe bodily functions are reflective of societal and medical views of those functions and, in turn, shape the way they are experienced on the individual level.[61] Terms like *symptoms, withdrawal, decline,* and *cessation* usually have negative connotations; Martin suggests that it would be helpful if we could develop a new vocabulary to describe menopause.

Health psychologist Paula Derry notes that menopause and senescence are often conflated so that any disease or disorder that happens to women several decades after menopause is referred to as being "postmenopausal."[62] For this reason, osteoporosis is often associated with menopause, even if it occurs two to three decades later. It would be equivalent to referring to health events of the 30s and 40s as "postmenarcheal" or postpubescent. Clearly this is true in one sense of the term, but it is not a useful designation for the same reasons that referring to something that happens to a woman in her 80s as "postmenopausal" would not be very meaningful.

Occasionally one will run across an article about "male menopause." The very concept of "male menopause" used to describe the psychological challenges some men face at midlife suggests that female menopause is all about psychological issues. Clearly men do not experience the hormonal and reproductive changes that occur in women at this time, nor do they cease menstruating, so the concept makes sense only if menopause is seen as a psychological problem. For men, the decline in reproductive function that occurs late in life is highly variable and cannot be measured by a simple concept as one year from a given event (the last menstrual period in women).

Even the medical view of menopause is highly variable across cultures. For example, in Japan, "menopausal syndrome" is recognized clinically, but it is seen not as a result of hormone decline, but as a result of moral decline that occurs in women who "have too much time on their hands" and focus on themselves rather than on their families and communities.[63] In a sense, physicians are blaming women for the ambivalence they feel when their children, in whom they are expected to invest so heavily, grow up and leave home, a time that usually coincides with menopause. The idea that a woman's identity is closely tied to her role as mother is not unique to Japanese culture, of course, and is commonly reported as a reason that menopause is stressful in Western societies.

Timing of menopause is subject to being defined as normal or abnormal. If menopause occurs before age 40, it is referred to as "premature ovarian failure" or POF. Obviously this somewhat arbitrary cutoff age would be meaningless in a non-contracepting population where women may become pregnant in their late 30s and never resume cycling after several years of lactation.

Whether perceived as a medical "problem" or not, the menopause transition is associated with a number of physiological changes that are experienced by women at this time and are usually explained by the changes taking place in ovarian function. Some of the most commonly reported sensations in the US include hot flashes, sweating, insomnia, vaginal and urinary discomfort, headaches, fatigue, irritability, depression, moodiness, and weight gain. Often these are divided into "physical symptoms" (hot flashes/sweats, and vaginal atrophy) and "psychological symptoms" (all others), and the frequency with which menopausal women report them is highly variable. Estrogen replacement therapy (ERT) has been found to alleviate hot flashes and vaginal discomfort, whereas there is no evidence of an effect on other symptoms.[64] One conclusion from this is that hot flashes and vaginal symptoms are related to the hormonal changes that accompany menopause, whereas the other symptoms co-occur with psychosocial changes many women experience during this time.[65] It has been estimated that more than half of all menopausal women in the United States experience hot flashes at some time during the perimenopause,[66] although they are reported more often in African American than in Euro-American women.[67] On the other hand, hot flashes are frequently reported in women 10 or more years before menopause and before significant changes in menstrual cycling.[68]

Hot Flashes and Night Sweats

Anthropologist Lynette Leidy Sievert assembled a table of "menopause complaints" reported by women from different populations around the world. Here are the ones listed most frequently: Philippines, Singapore, Taiwan—headache; Thailand— dizziness in one study, hot flashes in another; UK—depression; US (various populations)—aches and stiff joints or tenseness. Only two populations on this table listed hot flashes or night sweats in the top four complaints,[69] calling into question the idea that hot flashes and night sweats are expected to occur at menopause, are stressful, and can be explained on the basis of hormonal changes alone. It is clear from extensive cross-cultural research that hot flashes are not universal phenomena of menopause and are experienced and reported under conditions where environmental factors (climate and altitude), culture (marital status, religion, attitudes toward menopause and aging, diet, smoking, and reproductive history), and biology (genes, hormone levels, sweating patterns) all intersect and influence their expression (see Figure 9-1).

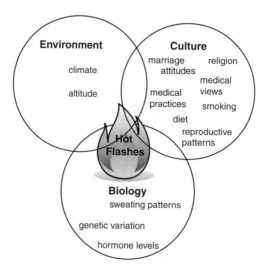

Figure 9-1 Culture, environment, and biology interact to influence hot flash expression. Modified from Sievert, 2006, page 141.

As with other menopausal complaints, the evolutionary medicine take on hot flashes calls into question whether or not they were commonly experienced in the past, and if they did occur, whether they were stressful or interfered with daily function. Recalling the unusually high levels of reproductive hormones that Western women have throughout their lives, it may be that withdrawal of these hormones has a greater impact on the vasomotor system than it does for women with lower levels and fewer lifetime menstrual cycles.[70] Perhaps a woman who starts menstruating at age 17, has 5-6 pregnancies, nurses for several years, and has only 60–100 menstrual cycles in her lifetime experiences fewer physiological disruptions at menopause than a woman who experiences 350–400 menstrual cycles with higher estrogen and progesterone levels.

Osteoporosis

Bone mineral density (BMD) declines at the time of menopause, a phenomenon that is supposed to be associated with declining estrogen. Because of this association, the medical view is that providing external sources of estrogen can reverse that trend and increase BMD. Not surprisingly, things are not that simple. In fact, BMD decline slows after about 5 years and there is evidence of what has been called "compensatory processes" whereby bones become wider to compensate for the decreased mineralization.[71] This is another situation in which context is important: bone mineral loss in the context of smoking, low levels of exercise, high levels of fat intake, high reproductive hormones throughout life, and some disease processes may be related to increased risk of fracture. In the absence of these other factors, however, estrogen decline and associated decrease in BMD may not have much of an impact on fracture risks. Evidence in support of this is seen in Mayan women who show decreased BMD following menopause but no increase in fractures.[72] This may be another example of an inappropriate designation of "normal." If "normal" mineralization is what pre-menopausal women have, then the decrease following menopause must be "abnormal" and warrant some sort of intervention. Certainly there are cases in which decreased BMD is associated with fractures, but in many instances, the fractures are more closely related to lifestyle factors (like smoking, obesity, low rates of exercise that improve balance) than bone mineral levels.

An old wives' tale is that each pregnancy costs a tooth, referring to the amount of calcium that is sometimes diverted from a woman's skeleton to her developing fetus. The ability to recruit calcium from the skeleton is advantageous for fetal development, but there is a cost later in life when a woman is at risk for developing osteoporosis.[73] The ovarian hormones work with other hormones to regulate calcium balance in bones, blood, tissues, and organs. Progesterone in particular seems to increase bone mass, probably by enhancing the ability to absorb and use calcium, which is especially important during the last trimester of pregnancy. More important than pregnancy for bone development is lactation, a period when infant access to calcium depends on the mother's calcium levels, and it may draw down her skeletal reserves even further. If the interval between lactation and the next pregnancy is too short, her skeletal reserves may not have time to recover. When the body allocates calcium, the needs of the fetus and nursing infant take priority over the mother's needs. In fact, long periods of lactation take a toll on a woman's bone mineral levels, especially if dietary sources of calcium are limited. If she breastfeeds right up until the time when reproductive hormones drop at the menopause, she may begin the postmenopausal period with a deficit.

When the reproductive hormones decrease at menopause, this triggers considerable changes in metabolism of calcium and bone. The systems that enabled a woman to absorb more calcium from the foods she consumes and maintain high levels no longer work well now that their job (reproduction) is over. This means that calcium levels depend to a greater extent on what can be taken from bone. Osteologist Alison Galloway suggests that the drop in estrogen and progesterone at menopause serves as a signal to release restrictions on mobilization of calcium from bone.[74] In her words, the body "borrows" calcium from the bone, but "this is a debt that cannot be repaid."[75] Again, this is the price we pay for a body "designed" for reproduction, not for long and healthy lives. Thinking about osteoporosis this way leads us to understand that it is the result of a very successful adaptation and not necessarily pathological. For women in health-rich populations, however, the drop in ovarian steroids is more precipitous because the premenopausal levels of these hormones are high; this may mean that the effects on bone demineralization are even greater and perhaps move us into the realm of pathology requiring medical intervention.

Sleeplessness

Sleep problems and insomnia are frequently reported by women during the menopause transition and often accompany aging in general. Women who report hot flashes and night sweats also report that these are disruptive of sleep, so the hormonal changes that seem to influence these symptoms may also underlie sleep problems. Unresolved, however, is which comes first, sleeplessness or hot flashes. In other words, does a hot flash wake you up or do you wake up and then have a hot flash? Most research supports a link between hot flashes and disturbed sleep, although the direction of the association is not fully understood. Furthermore, sleeplessness may also underlie mood disorders reported by women experiencing menopause, which may explain why mood disorders and hot flashes are associated in many studies. In 1977, two researchers suggested that estrogen therapy might decrease both "psychological symptoms" and insomnia by relieving hot flashes, and referred to this possibility as a domino effect.[76] In the present interpretation of the domino effect, hot flashes (perhaps caused by changes in estrogen level) disturb sleep, and sleep disruption causes mood problems.

In our research using daily reports of sensations,[77] events, and mood, my colleagues and I were able to test for the sequence of sensations as a preliminary investigation of what comes first in the domino hypothesis. We found that hot flashes on one day were followed by reports of sleep problems, which in turn, were followed by mood disturbances. But that is apparently not the whole story. When we controlled for sleep problems, the predictive value of hot flashes on mood was not eliminated, supporting an independent effect of symptoms. Further, having sleep problems was a better predictor of mood than were hot flashes. Thus, hot flashes and sleep problems likely work both together and independently to influence mood at the time of menopause.[78]

Sleep itself can be viewed from an evolutionary perspective. There are clearly internal mechanisms that regulate sleep, but they can be easily disrupted by both internal and external events.[79] In fact, "normal" sleep is, like many other aspects of our lives, culturally constructed and varies widely across cultures. For many of us, a good night's sleep happens for about eight hours in a comfortable bed in a quiet, dark, slightly cool room. Ideally, we are virtually dead to the world and alone or with our mate or infant. If we awaken frequently during the night because of noises, temperature changes, movements of bed partners, or internal states, we complain that we did not have a good night's sleep. We are advised to go to bed at a set hour each night and wake at a set hour each morning. But consider the circumstances under which sleep took place in the past. Clearly some level of sleep is necessary for healthy physical and mental functioning (for instance, memories are formed during sleep), but there are risks from sleeping too soundly. Predators, enemies, and fires going out probably all favored the ability to awaken quickly in order to respond to challenges. Sleeping close to others provided both warmth and a level of protection for adults as well as for infants (see Chapter 8). Although most sleep took place at night (we are diurnal animals, after all), it is highly unlikely that it took place between set hours. In Carol Worthman's words, sleep was "fluid" and occurred within "fuzzy boundaries in time."[80]

Furthermore, Worthman suggests that one of the reasons that some of us have more difficulty sleeping than others is that we minimize potential distractions, making it too quiet and giving us too much time to think and worry. Human sleep evolved in a context that included crackling fires, quietly breathing family members, and "munching of animals"[81] that gave cues of security and enabled relaxation of vigilance.[82] She refers to this as a "cue-dependency model of sleep" and suggests that without these cues we turn to other indicators of security, finding with our worried thoughts very little relief from cognitive stresses that disrupt sleep.[83] Maybe our teens who like to sleep with the radio on are onto something.

As we age, we find that we have more difficulty falling asleep and staying asleep. For the elderly, in particular, sleep problems are quite common. In fact, sleep aids and even powerful sleeping pills are often resorted to by people who have difficulty falling asleep or find themselves waking frequently in the night. Breathing interruptions known as sleep apneas become increasingly common with age, so some of the arousals may be adaptive in that by awakening, a person triggers a return to normal breathing. In cases like this, a person who takes a sleeping medication may find that he or she fails to awaken when respiration falters, leading to death. The evolutionary medicine view of frequent awakenings in old age is that they often serve the purpose of re-establishing respiration and are thus advantageous, defenses rather than defects.

Because sleep is believed to be as important to human health as food, difficulties with sleep, including insomnia, have spawned a slew of treatments and medical specialties in the US. On the other hand, it is unclear whether sleeplessness has significant costs in terms of morbidity and mortality,[84] and it appears to be relatively easy to recover from. That does not keep people from fretting about it, however, and one of the most powerful sleep disrupters is anxiety, including anxiety about getting enough sleep. Given the fragility of sleep and the ease with which it can be disrupted (adaptive in the past as part of a "well-honed capacity for vigilance"[85]), it is not surprising that the physiological and psychological changes experienced by many women at menopause often lead to altered sleep patterns.

Other Menopausal "Complaints"

Vaginal dryness is another common complaint of menopause and it is likely due to underlying hormonal changes. For women who are not sexually active, however, they may not notice vaginal changes. In many societies today, including foraging societies, women marry men much older than themselves, meaning that by the time they reach menopause and the associated vaginal dryness, their husbands are already dead and their level of sexual activity has decreased or stopped altogether. Of course, women may be sexually involved with men other than their husbands, but although there are anthropological reports of sexually active postmenopausal women,[86] other evidence suggests that it is not very common among older women, even where it is reported and joked about.[87] Studies of women in industrialized nations show evidence of both increased and decreased sexual activity and libido following menopause.[88]

A common complaint of menopausal women in health-rich populations is the associated weight gain that often occurs after menopause. Up to a point (say about 5 pounds), this too may be advantageous for health. Estrogen is stored in fat tissue and androgens can be converted to estrogen in fat. Women with lots of body fat produce more estrogen post menopause,[89] but postmenopausal obesity increases the odds of getting breast cancer.[90] It seems reasonable to argue that putting on a little extra weight (probably no more than 5 lb) at menopause may be advantageous for bone health. So rather than curse those few extra pounds, perhaps they should be worn proudly as evidence that our bodies can make up for decreases in estrogen that occur when reproductive years end. Obviously, too much of this "good thing" can tip the scales to compromised health.

Why Is Menopause Stressful?

The short answer to this question is that it is not stressful for most women in the world, as far as can be ascertained by social scientists. Like so many other psychosocial stresses, even when there is an underlying biology to explain them, how the changes are perceived is highly dependent on context. Furthermore, whether or not menopause is welcomed or dreaded may also vary with context.[91] It can be predicted that in cultures where women's status rises at menopause, the hormonal and other changes that occur at this time are not perceived as stressful and may not even be noticed, or may be welcomed. In the United States, youth is so highly valued that when a person begins to exhibit visible signs of aging, their status often declines, especially for women. Menopause is thus, for many women, a clear sign that they have achieved a lower status than they may have had when they were younger. Where status declines with menopause, it is not surprising to find many women reporting the time as emotionally and physically taxing.

In a review of midlife women (referred to as those "in their prime") across world cultures, anthropologist Judith Brown notes that in contrast to most life phase events (like puberty and childbirth), menopause is not usually celebrated ritually and is rarely reported in anthropological writings.[92] This is further evidence that it is a "nonevent" in many populations because there is no specific biological marker that works when menstrual cycling is not a phenomenon that marks a woman's reproductive years in the same way it does in populations in which women use contraceptives. In other words, the actual transition is not remarkable and may go unnoticed not only by the woman experiencing it but by the community of which she is a part. Once it is clear that they are beyond their childbearing years, however, many women do experience status changes that include removal of restrictions typically placed on women of reproductive age, increased decision-making authority, and even a new, specialized status (like midwife or ritual specialist). Among the !Kung, anthropologist Richard Lee reports that postmenopausal women become much more sexually adventurous, their dress becomes much more provocative, and they often take younger men as lovers.[93] On the other hand, in India where female sexuality is linked with blood, postmenopausal women are not seen as sexual beings and are not expected to be sexually active.[94]

Anthropologists Jane Lancaster and Barbara King propose that modern women may experience the end of ovulation under a different hormonal milieu from that of our ancestors and women in "natural fertility" populations today.[95] For example, !Kung women reportedly breastfeed their last infant for a longer time than for previous infants and when they terminate breastfeeding, perhaps in their mid-40s, they simply do not resume ovulation. Thus, the hormonal milieu of menopause includes the hormones of lactation (especially prolactin and oxytocin), which may modulate the effects of other changes. We noted in Chapter 2 that women in health-rich populations have higher circulating ovarian hormones during the menstrual cycle and pregnancy than do women in health-poor populations. It also seems that the former report more stress at menopause and more symptoms than the latter, and higher estradiol levels may exacerbate physical changes at menopause for Western women.[96]

The perspective of evolutionary medicine would argue that "returning" to the lifestyles under which human reproduction evolved would alleviate some of the problems that women face today with menopause. Without taking the recommendation to extremes (no birth control, self-imposed undernutrition), it seems possible that by increasing exercise, reducing fat intake, and minimizing exposure to industrial biochemicals, women would experience fewer problems at menopause. In this way, adopting the behaviors and diets of our ancestors may go a long way toward not only reducing the stresses of menopause but prolonging healthy lives as well. We will see in the next chapter that even more unusual than menopause is the phenomenon of women living several decades beyond the age of reproduction.

10

What Good Are Old Women?
Quite a Lot, Thank You

How long did people live in the past? Studies of human skeletal remains are not very helpful because estimating age beyond about 45 is difficult. But studies of recent foragers belie the argument that only a few people lived beyond age 50. Not only is it common to find people in their 50s and 60s among foraging populations like the !Kung, Ache, and Hadza, but many of these elders maintain physical vigor equivalent to or better than their peers in industrialized populations.[1] It is true that mortality at younger ages is higher than in modern populations, so life expectancy at birth or an average life span is lower than 50. But once a woman reaches age 45, close to the end of her reproductive years, she can expect to live about 20 more years.[1]

Grandmothers and Reproductive Success

Most long-lived, group-living mammals have in their social groups as many as three generations present at any one time. Examples include elephants, whales, and many primates. For primates who live in matrilocal groups, that usually means three generations of females: infants, their mothers, and their grandmothers. A famous example comes from Jane Goodall's studies of a Tanzanian chimpanzee social group in which Flo, her adult sons Faben and Figan, and her daughter Fifi lived together. Flo was a high-ranking female and her presence had a number of positive effects on her offspring. For example, Fifi was able to stay in the troop into which she was born, whereas the more typical pattern among chimpanzees appears to be for young females to leave their birth troops at maturity. By staying with her mother, Fifi was also able to rise to a high status; she began reproducing much earlier than most chimpanzee females and not only set the record for reproductive success at Gombe, but one of her sons became the largest male ever recorded at Gombe. Two of Fifi's sons rose to high status in the dominance hierarchy and her daughter began reproducing even earlier than Fifi did. There is little doubt that grandmother Flo's status had an effect on her daughter's (and thus her own) reproductive success.[2] There is no evidence, however, that Flo contributed directly to the care and feeding of her grandchildren, although it is true that she was not in good health at the time Fifi's first infant was born in 1971.

Anthropologist Sarah Hrdy notes that despite her reproductive success, Flo serves as a good example of why having offspring at later ages may not be a good way to achieve this success or why "stopping early" might be selectively advantageous. Flo reproduced for the last time when she was very old and in poor health, but that infant did not live long. Goodall proposes that this last pregnancy was so draining for her that she was unable to mother her other young offspring, Flint, and when Flo died, Flint died also, even though he was at an age when he should have been able to survive on his own. In fact, if Flo had stopped reproducing after Flint, he probably would have lived, perhaps going on to sire other offspring and increasing Flo's reproductive fitness through her grandchildren.[3]

Similar evidence that the presence of grandmothers has positive effects on reproductive success comes from observations of a number of other primate species. Again, it is not usually resources and direct care that older female grandmothers provide; rather, they help to defend the infants from other troop members (including infanticidal males) whose behaviors endanger them. In fact, observers report that grandmothers will often act even more vigorously in defense of infants than younger kin.[4] Grandmother Japanese macaques make a significant difference in survival of their grandchildren through the first year of life. Furthermore, females have much greater reproductive success if they have living mothers, even when those older females are still reproducing.[5] Similar reports have come from studies of vervets, langurs, and rhesus monkeys, as well as elephants. On the other hand, African lions and olive baboons, while showing extensive caretaking by adults other than the mother (known as allomaternal care), do not seem to have their reproductive success influenced by the presence of grandmothers.

These descriptions of primate social groups with three generations of females are not very different from what is seen in traditional human societies and even in extended family households in health-rich nations like the United States. What is different, however, is that in most cases the grandmother is not only helping her own older children but she also provides care and resources to her grandchildren.

Another view of menopause focuses not on the mother and early termination of reproduction (the "long-lived mother hypothesis") but on the grandmother who maintains health long after ceasing to reproduce. Known as the "grandmother hypothesis," this proposal assumes that termination of fertility at about age 50 is a given, but that natural selection favored a long postreproductive period in women's lives because by ceasing to bear and raise their own children, postmenopausal women would be freed to provide high-quality care for their grandchildren.[6] In this scenario, older women "trade" their diminished chances of successfully raising an infant for enhanced opportunities to help raise their grandchildren. This is simply the continuation of a behavior that women have practiced for most of their adult lives: providing food and care for children who have been weaned but who are not yet capable of getting their own foods in sufficient quantity and quality to survive. This continuity-of-care hypothesis also explains why so much of the focus on older people as alloparents is on grandmothers.

When the grandmother hypothesis was first proposed by Kristen Hawkes and her colleagues, it included supporting evidence based on their studies of the Hadza, a foraging population in Tanzania.[7] Among these people, when a woman gives birth, her time providing food for her older children is severely curtailed and remains lower than usual for several months. During this time, the grandmother increases her foraging to make up for the reduction by the new mother. Certainly her success is increased if she is still in good health and able to travel widely gathering food. Thus, the argument is that natural selection not only favors termination of reproducing at about age 50 and provisioning of older infants by grandmothers, but it also favors continued vigor and good health in the grandmothers until their own daughters cease reproducing and become provisioning grandmothers themselves. Notably, this argument also proposes that matrilineal proximity would be favored as well, calling into question the assumption that early human social groups were patrilocal and that females dispersed at maturity. Older women who provision their sons' children would also increase their fitness,[8] although certainty of kinship is higher through matrilines than through patrilines.

Provisioning by postmenopausal women of the offspring of any of their kin would be selected, a phenomenon noted for the Hadza, for whom the most important component for increased offspring survival was a postmenopausal helper, no matter how she was related to the offspring. In other words, when the word "grandmother" is used in this scenario for the evolution of postreproductive healthy lives, it is not necessary to invoke the literal meaning of grandmother. Rather, a woman (or, indeed, a man) of grandmotherly age who provides resources to youngsters with whom she shares any genes, including those of her sons, would enhance her own fitness. Obviously the payoff would be greater for literal grandchildren, but as long as the provisioning is not indiscriminate, the model holds up.[9]

If grandmothering is related to enhanced reproductive success, one would expect that on average, women with long postreproductive lives would have more grandchildren. Since the initial proposals of a grandmother effect on reproductive success, a number of demographic and ethnographic studies have confirmed it.[10] An unusual set of data from Gambia that covered the years 1950–1975 was used to assess the effects of grandmothers on infant survival, yielding evidence that infant mortality was significantly reduced between ages 1 and 2 years for infants who had living grandmothers. Other relatives (fathers, grandfathers, siblings) had no effect on infant survival and grandmothers had no effect for infants under 1 year.[11]

Analysis of demographic records from Canada and Finland for the 18th and 19th centuries, when grandparents were integral to the success of rural families, shows that women with long postreproductive lives did indeed have more grandchildren, but that the benefits of grandmothers diminished when their own daughters ceased reproducing.[12] Grandmothers were particularly effective at enhancing survival of their grandchildren up until they were 60 years of age. Their presence made no difference to the health of infants under 2 who were being nursed by their mothers, but the grandmother made a big difference when infants were over 2. In contrast, another study of the effects of grandmothers on infant survival in Germany in the 18th and 19th centuries demonstrated a positive effect of maternal grandmothers on infant survival between ages 6 and 12 months but a negative effect of paternal grandmothers.[13]

Finally, there is at least one report of a negative effect of grandmother's longevity on daughter's fertility, but a strong effect of a mother's longevity on her own fertility.[14] This seems to support the long-lived mother over the long-lived grandmother hypothesis, but it also suggests that there is likely a great deal of variability across cultures in the quantity and quality of grandmothering, so conflicting results of the effects of grandmothers may be due to some of this cultural variation. This is, of course, not surprising to many anthropologists who argue that it is extremely difficult (if not impossible) to test evolutionary hypotheses on human populations because of the high degree of variability in almost every known behavior.

Longevity

Humans are an unusually long-lived species, with a maximum life span potential estimated to be about 120 years. This maximum probably has not changed in the last several thousand years, but life expectancy at birth (the average length of life) has certainly increased in the past several decades because of advances in standards of living, hygiene, and medical care, most especially the treatment and prevention of infectious diseases. It can be argued that humans age throughout their lives,[15] but in the colloquial sense, aging is equivalent to senescence, the physiological decline in all systems of the body that occurs toward the end of life. Throughout adulthood, there is a gradual decline in muscle mass and strength, in bone mineral density, in immune function, and in our ability to synthesize proteins. Associated with this decline is an increased risk for the chronic degenerative diseases that find their places on death certificates in health-rich nations.

I have discussed the importance of long-lived grandmothers[16] for survival of children, but longevity in general appears to be advantageous to a species that depends so heavily on information handed down through the generations; in other words, for a species so dependent on culture. Is the unusual longevity seen in humans associated with the expansion of culture? Anthropologists Rachel Caspari and Sang-Hee Lee argue that it is, and that longevity appeared as recently as the Upper Paleolithic, 30,000 years ago.[17] They base their argument on dental evidence from 768 hominid fossils ranging from early australopithecines to Upper Paleolithic Europeans. Their analysis led them to conclude that longevity increased significantly throughout the course of human evolution, with the greatest increase seen in the most modern sample, that from the Upper Paleolithic. In this population there were more older individuals than younger ones for the first time. They further argue that this was the point at which older people were increasingly important for the transfer of specialized knowledge and skills to younger generations. Finally, they suggest that a demographic shift occurred at this time, which led to population expansion as fertility increased with longevity and helping grandmothers. This increase in longevity may also explain the expansion of culture and creativity seen in the Upper Paleolithic.[18]

Given the evidence that healthy long-lived grandmothers contribute to the survival and reproduction of their offspring and grand-offspring, it can be argued that genes for long lives were disproportionately passed to the descendents of these grandmothers. A great deal of research suggests that genes are involved in longevity,[19] and certainly there is variation around a mean expected age of death. Thus, throughout history, the average age of death for humans gradually increased so that it is several decades higher than for almost all other mammals. Today, medical science keeps people alive and healthier longer than they would have lived in its absence, but that aspect is due more to culture than to biology. We can conclude, therefore, that human longevity owes its characteristics to both biology and culture and how those interact in the individual.

Most animals die from what can be called "extrinsic" causes (such as predation, starvation, cold, heat, drowning, and disease), whereas modern humans most often die from "intrinsic" causes (accumulation of cellular and molecular damage) because they are "protected" from the agents that killed their ancestors. Our bodies have ways of repairing cellular damage from the moment it occurs, even in utero, but eventually this machinery wears out and is no longer able to perform its function, resulting in accumulations of damage that lead ultimately to death. The repair machines are likely subject to selective forces and result in the great variability seen in the healthy aged in modern populations. Furthermore, the success of those machines in repairing damage is heavily influenced by environmental and lifestyle factors, which helps to explain greater longevity in health-rich than in health-poor populations. As an example, there is an enzyme that is responsible for DNA damage repair that exhibits very high activity levels in centenarians.[20]

Another enzyme that seems to be related to longevity is telomerase, which protects the DNA sequence at the end of each chromosome, an area known as the telomere. Each time a cell divides, the telomeres are shortened, eventually reaching the point at which they can no longer divide and are unable to maintain healthy tissues and organs. In the laboratory, the enzyme telomerase can lengthen telomeres, allowing the cell to continue to divide. But this may not be a good thing in a person, since the only cells that can divide indefinitely are cancer cells. Although this research is not likely to lead to a lengthening of the life span, it may contribute to a better understanding of cellular functions and cancer.

Another anti-aging factor that eventually wears down is one responsible for removing or repairing damaged proteins. Furthermore, an accumulation of damaged protein after long life contributes to cataracts, Alzheimer's, and Parkinson's disease.[21] All of these processes that begin to lose their functioning ability at the end of life are important for maintaining health early in life and appear to be heritable. As is often said, if you want to live a long and healthy life, choose your parents well.

Another reason that humans die less often from extrinsic causes than other mammals is that we have big brains that we use to think our way out of dangerous situations and learn what plants may make us sick or kill us. This may be one of the explanations of the link between large brain size and longevity: selection favored big brains because they allowed cognitive skills that were favorable to survival (and reproductive success), but in order for them to be very useful, people had to survive long enough to learn the important skills and gain the knowledge.[22] (If that sounds circular it is; much of evolutionary history reflects feedback mechanisms operating through time that look circular.) This, in turn, allowed for long childhoods for learning and intergenerational transfers of knowledge and resources.[23] In fact, Ronald Lee, an economist, has proposed that the flow of resources transferred from one generation to the next can alone explain the long period of postreproductive survival in humans.[24] In his view, successful reproduction (fitness) is not just measured in the number of genes that are passed on from one generation to the next; equal in importance (or more important for a species like humans) are the resources like food and child care that are passed on. This proposal has come to be called the "intergenerational transfer hypothesis," and I will elaborate upon it further in the next section.

Health between 50 and 70

If the grandmother and the intergenerational transfer hypotheses about longevity are correct, then along with staying alive, older people need to be mentally and physically healthy so that they can contribute to their families and communities rather than drain them of resources as dependents. Among some foraging populations, net productivity is high from early adulthood until it reaches a peak about age 45. Then it declines gradually but still remains positive until about age 60,[25] a profile quite different from that of the chimpanzees. The age profile of productivity reflects physical abilities, skills, and knowledge and begins to drop off as these wane with health late in life. In particular, hunting (mostly by males) and digging for roots (mostly by females) are negatively affected by health after age 60. An important link in all of this is that humans, differing from other primates, consume foods that are often hard to get, may require processing, and tend to be of higher quality[21]—in other words, a diet that requires intelligence. Furthermore, from an evolutionary standpoint, there is evidence that the advantages for longevity of grandmotherhood decrease precipitously by age 65.[26] This is probably the point at which an older person, for health reasons, is no longer contributing at a level consistent with enhancing family reproductive success. It is also the point at which a child born to a woman even as old as 50 would be independent.

Brain Aging

As I have emphasized throughout this book, brains are very expensive to maintain, so much so that large brain size and associated intelligence must have been very important in human evolutionary history for our species to have afforded something so extravagant and outsized. Furthermore, the normal human brain retains full functioning until at least age 60 in modern humans,[27] after which time structural decline typically occurs, although there is great variation among individuals with regard to how fast this decline occurs or how much it affects cognition.

As important as intelligence is to human success, it is curious that cognitive decline is so strongly correlated with aging. Humans are not unique in this, and cognitive decline has been noted for many species that live long lives. In fact, humans show later onset of brain aging than monkeys and apes, which may be related to the long period of infant dependency necessitating care from a physically and mentally healthy mother for at least 10–12 years. Reproductive hormones may be protective against cognitive aging and serve to facilitate adequate provisioning and caretaking by parents in those dozen years after the birth of the last child.[28] When reproductive hormones decline in women following menopause, the protective factor is gradually withdrawn, providing a rationale for why hormone replacement therapy (HRT) appears to delay the onset of cognitive decline associated with such disorders as Alzheimer's disease. This is another phenomenon in which selection for survival at early ages may become negative for health in later years.

The gene that codes for the fat-transporting protein apolipoprotein E (apoE) may be an example of one that has differential effects at different phases of the life span,[29] although the early life advantages are still not clear. The gene has several variants (alleles), one of which, apoE ε4, is associated with increased risk for Alzheimer's and cardiovascular disease. (In fact, apoE ε4 is the single most common risk factor for Alzheimer's.[30]) Two other variants (ε2 and ε3) appear to be protective against both. Most interesting from the perspective of evolution is that other primates appear to have only the apoE ε4 variant (more technically, apoE ε4-like, because they are not exactly the same). Perhaps the two new variants are responsible for our lengthened reproductive lives and overall longevity and they may have appeared after the human and ape lines diverged.[31] They could have increased in frequency because parents and grandparents who had one of the recently evolved alleles were mentally and physically healthier late in life and could provide better care for their offspring, passing along more copies of the alleles that are protective against brain and cardiovascular decline. Maybe these two variants "evolved to slow down both brain and cardiovascular aging."[32]

Genetic surveys support the hypothesis that apoE ε4 is the ancestral allele and that apoE ε3 has increased in frequency over the course of the last 200,000 years.[33]

They also confirm that there is a great deal of population variation in the frequencies of the allele (with apoE ε3 ranging from 65% to 85%), which may underlie some of the variation in prevalence of cardiovascular and Alzheimer's disease. Countering the disadvantages of the apoE ε4 allele at older ages, there must be advantages earlier in life, for example, in protecting against infectious diseases that affect the young.[34] Indeed, ApoE ε4 carriers seem to be better able to mount an inflammatory response to pathogens and other stresses, adding to its value in earlier years.[35] In sum, apoE ε4 appears to be "good for" young people because it provides protection against infectious agents and enables them to survive the reproductive years. ApoE ε3, on the other hand, is good for older people (ages 40–70) because it provides protection against chronic diseases of aging and enables them to contribute to the well-being of their offspring and grand-offspring. In environments with high levels of infection today, people with the apoE ε4 allele are advantaged but in environments with low levels of infection, the healthiest and longest lived are those with the apoE ε3 allele.[36]

Individuals with apoE ε4 alleles are particularly susceptible to diseases caused by a small bacterium known as *Chlamydia pneumoniae*, the primary cause of pneumonia. Biologist Paul Ewald and others argue that this bacterium is implicated in dozens of diseases ranging from acute respiratory diseases to chronic diseases like atherosclerosis.[37] The mechanism appears to be that the bacterium attaches easily to the apoE ε4 protein and moves with it into cells. In their view, the cells it enters have an effect on the type of disease that develops: in brain cells, it may develop into Alzheimer's; in arteries it is manifested as atherosclerosis. Of course genetic and lifestyle factors also play roles in the development of these diseases, but they provide a rationale for considering all three factors (genes, lifestyle, pathogens) for most chronic diseases, as Ewald suggests. In the case of apoE ε4, he argues that it is the "Achilles heel" that makes a person more susceptible to *Chlamydia pneumoniae* infection, but how it plays out in the body may vary with other factors, many of which become more relevant with aging.

Another interesting thing about the apoE allele is its relation to diet. As is well known by now, humans descended from animals that were primarily fruit and plant eaters, consuming animal protein only occasionally. A dietary change that seems to have had wide-ranging impact on hominin evolutionary history was an increase in animal consumption, possibly associated with brain expansion about 2 million years ago. We recognize today that high-meat diets are related to high blood cholesterol with its associated higher risk of various chronic diseases and disorders, so it is not immediately clear what meat eating has to do with longevity. One proposal is that increased consumption of animal protein through time was associated with selection for "meat-adaptive genes" that are protective against the expected negative consequences of meat-rich diets.[38] The candidate for this meat-adaptive gene is the apoE ε3 allele. In fact, chimpanzees with their apoE ε4 alleles are highly susceptible to hypercholesterolemia and cardiovascular diseases when they are fed diets high in animal protein and fats in captivity, providing further evidence of an interactive effect between meat consumption and high-cholesterol-linked diseases.

Mental Health

There is little doubt that people cannot learn as well or remember as easily in their 80s as they can in their 20s, but aspects of cognition vary in the extent to which they decline and how long they persist. One important skill that improves with aging is social and emotional functioning, and people tend to become more satisfied with their social and subjective lives.[39] The phrase "paradox of aging" refers to the phenomenon that one's sense of well-being increases even as physical and some aspects of cognitive functioning decline. An evolutionary view suggests that this is not a paradox at all but instead is part of an evolved psychosocial system that enhances reproductive success via positive and nurturing social interactions, especially with kin. In other words, older people who feel good about themselves and their families are more likely to provide care and resources to their descendents, who, with their help, are more likely to survive and reproduce. Some scholars[40] suggest that associated with this is the ability to monitor the amount of time left to live and that when people perceive that this is decreasing, their "motivation shifts from goals related to personal advancement to goals that benefit others."[41] They further argue that this ability "to monitor place in the life cycle" is unique to humans and was selected because it increases the likelihood that grandparenting behaviors will emerge as people age. Interestingly, this phenomenon may account for the commonly expressed need to be close to family that followed upon the events of September 11, 2001, in the United States when more than 3,000 people were killed as planes flown by terrorists slammed into the World Trade Center (New York) and the Pentagon (Washington, D.C.). It also may explain the lower rates of depression and anxiety in older people.

Social networks tend to decrease with aging. Certainly this may be due in part to higher rates of death among peers as people age, but there is evidence that older people make conscious decisions to reduce their networks to include only those who are emotionally significant, predominantly those in the kin group.[42] The theory underlying these findings is known as "socioemotional selectivity theory" and it supports the proposal that inclusive fitness is promoted when people shift toward investment of time and resources in kin—in other words, when they adopt grandparenting behaviors. One could argue that this theory underlies intergenerational transfers and increasing longevity in humans. Further, "older adults for whom the association between limited time perspective and selective care for familiar individuals was stronger would have contributed more to the fitness of their descendants."[43] Moreover, it appears that older people retain information that is especially important for group survival in stressful times, such as the ability to "see the big picture," that may have value equal to or greater than the new knowledge that younger adults can more readily attain. It is no coincidence that the best storytellers and the best negotiators are the oldsters.

One study confirmed that emotional well-being improves from age 50 to 80; it demonstrated with MRIs that perceptions of positive emotions improve, and control over negative emotions becomes more refined, with a resulting heightened sense of well-being.[44] Thus with aging may come an understanding to not "sweat the small stuff" and "live one day at a time," both of which probably make life more pleasant, in contrast to the views of grumpy old men and women. This evidence that emotional cognition and behavior improve with age is welcome news as we face a future when as many as one third of the world's population is predicted to be above the age of 60.

As we respond to renewed calls to service following the 2008 presidential elections in the United States, it may be that oldsters are the best candidates because of their available time, perspectives, and well-developed socioemotional cognitive abilities. Too much of a good thing may cause problems, however. The relationship between volunteering and well-being appears to be U-shaped: well-being was lowest at both low and high levels of volunteer activity.[45]

Are Grandmothers Healthier?

Most of what I have been discussing so far is how postmenopausal women enhance their reproductive success through raising their own young children or helping their daughters and sons raise their grandchildren. But other than reproductive success, is there anything in it for the older women? There is evidence that women who have grandchildren live longer than those who do not, but are they healthier? It is one thing to live a long life, but equally or more important for provisioning or active child care is how healthy a woman is during her later years. Looking for proximate explanations of child care by grandmothers, is there any evidence that being around young children enhances one's own health?

It seems logical to claim that the health of older women is better with cessation of reproduction than it would be if reproduction continued. Given the challenges of pregnancy, birth, and caring for a very young infant, it is highly likely that death would have been directly due to reproductive factors if women continued to ovulate and get pregnant into old age. It seems silly to put it this way, but surely post-menopausal women in the past were healthier than they would have been with continuing risk of pregnancies (the same would be true today, but mechanisms of birth control are available to keep pregnancy from happening).

For a grandmother to have an impact on survival of her grandchildren through her own work and caretaking efforts, she herself needs to be healthy and capable of productive work and caretaking. So it is not just that selection would act on grandparent longevity[46] but also on health and well-being in the postreproductive years. Furthermore, for there to be selection on grandmother caretaking, there must be proximate rewards that encourage continuing efforts. For example, providing nourishment for infants through breastfeeding has been selected in human females because it decreases infant mortality and increases a woman's reproductive success. Fortunately, for most women, breastfeeding an infant is a pleasurable sensation,[47] so the behavior is positively reinforced by proximate mechanisms. As discussed in Chapter 7, breastfeeding a child has a positive impact on later health of the mother.

By the same reasoning, if caring for grandchildren or providing resources for them does not bring some pleasure to the grandmother, then she may *choose* not to continue or she may provide less-than-optimal care. Alternatively, if caring for grandchildren or providing resources compromises a grandmother's health, then she may not be *able* to continue. Clearly grandmothers[48] who feel good about caring for their grandchildren and who remain mentally and physically healthy through the caretaking years are likely to have the greatest positive impacts on infant and child survival. It is one thing to be physically and mentally capable of providing child care and resources, but equally important is having the motivation to do so.

In the United States, grandparents, particularly grandmothers, have become the focus of a great deal of attention from social scientists, health care workers, and even politicians because of the increasing numbers of older men and women providing exclusive or near-exclusive care for their grandchildren. In 1970, 2.2 million children (3.2% of all children) were raised in families headed by a grandparent; by 2000 that number had more than doubled, increasing to 4.5 million (6.3% of all children).[49] Most studies reveal that providing full custodial care is highly stressful and challenges the health of older women.[50] In fact, most studies of grandparenting proceed with the expectation that it has negative impact.[51] Full grandparental custodianship, particularly in social isolation, is likely a relatively recent phenomenon. In the evolutionary past if circumstances required that a grandmother assume full care of a grandchild, there were likely many other adults (alloparents) in the social group who could assist, so the burdens of child care were more widely distributed. In the United States today, custodial grandmothers are often socially isolated, providing one source for stress and compromised health.

This suggests that there may be an optimal level of grandmothering that has been selectively favored, but when the level of effort exceeds the optimum, the proximate rewards and overall health of the grandmother decrease. Furthermore, when the burden of care falls exclusively on the grandmother (as in custodianships), the health of the grandchildren appears to be compromised as well, in comparison to other child care situations (especially with two-parent homes).[52]

So if grandmothers have been selected to provide care for their grandchildren, at what point does the level of care become a burden and fail to enhance reproductive success? On the other hand, is there a level of grandmothering *below* which health and well-being are compromised? For example, are grandmothers who are estranged from their grandchildren or otherwise rarely see them less healthy and more likely to be depressed than custodial or day-care grandmothers? Perhaps a certain amount of contact with grandchildren is better than no contact.

It is worth mentioning that the causal arrow among the variables of grandmother caretaking, longevity, health and well-being, and reproductive success may be different from that proffered in the traditional view of the grandmother hypothesis (see Figure 10-1). As noted earlier, in that view the long-lived grandmothers who provide care and resources to their grandchildren would have greater reproductive success and therefore, genes for longevity would be favored. It is possible

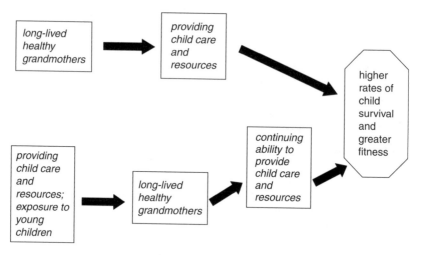

Figure 10-1 Effects of grandmorthers on child health and survival.

that child contact itself provides proximate advantages to health, well-being, and longevity so that the causal arrow can be reversed. In this case, it would not be necessary to have child care activities concentrated on biological kin. Just being around young children could render positive benefits, although if this relationship was established in the past, it is likely that all children in a social group were related to some degree to the older women in the group.

Anthropologist Cheryl Sorenson Jamison and her colleagues note that in the population they studied, grandmothers were 40% less likely to die in a given year than equivalently aged women (up to age 65, as noted earlier), suggesting a protective effect of being a grandmother. They suggest that among these effects is that if a woman is motivated to provide food for her grandchildren, she is probably also getting adequate nutrition for herself and may be healthier. They refer to this as a positive feedback mechanism whereby "the more the lives of the old are benefited, the more available they are to help the younger members of the group."[53] Furthermore, as we know from contemporary research on healthy aging, women who are providing a service usually feel better about themselves and are healthier.

Why Does Our Health Decline as We Age?

No matter how great our diet and exercise regimen has been throughout our lives, no matter how good our genes are, no matter how rarely we have been sick and how good our health care has been, our bodies still begin to fall apart soon after we turn 70. Why do we age at all? Historically, there have been two theories for aging related to evolution. One is that because natural selection acts most powerfully before the end of the reproductive years, mutations that accumulate after that point and cause ill health are not selected against. Related to this is the argument from pleiotropy that selection is simply more powerful during the reproductive years, so that genes that enhance reproductive success will be selected even if they mean poor health in the later years. Consider, for example, a gene that is neutral or even negative for reproductive success but positive for great health in old age. Compare that with a gene that is positive for reproductive success and simply terrible for later age health. Which will be passed down through the generations? Here is an interesting thought game about pleiotropy: a person might be willing to take a drug that will kill him in 20 years if it means he will not die tomorrow.[54]

People are living longer because they either survive or do not get the infectious diseases that have killed humans throughout recent history. But even if a person survives pathogen exposures, the very fact that infectious agents that cause inflammation are part of an individual's history may compromise later life health. One group of researchers examined levels of inflammation in what they called "two epidemiological worlds," that of a contemporary U.S. sample and that of an Amazonian tribe, the Tsimane.[55] The Tsimane live in an environment with high levels of parasites and other pathogens and have high mortality, with an average age at death of about 43 years. Blood tests done on these people also show much higher levels of inflammation than observed in U.S. populations, even when they showed no clinical signs of disease. This inflammation is associated with higher risk of cardiovascular disease. For populations living in low-pathogen environments, like the United States, inflammation is lower throughout life, aging is slowed, and lives are longer. Thus, decreasing the impact of infectious diseases has a much wider impact on contemporary population demography than previously thought. Inflammation is linked increasingly to a myriad of chronic diseases and disorders including type 2 diabetes, congestive heart failure, and Alzheimer's. Inflammation is good because it is the body's first line of defense against infectious agents and it starts the process of healing following injury. In fact, people with insufficient inflammatory responses would not live very long. But like so many other factors discussed in this book, too much of a good thing can be problematic, especially if it occurs early in life or is chronic throughout life. It appears that one of the hallmarks of a long and healthy life is low lifetime levels of inflammation.

How to Live a Long and Healthy Life

Eat lots of vegetables, fruits, grains, and foods high in omega-3 fatty acids, especially fish. Get a good night's sleep as often as you can but do not obsess about sleep to the extent that you lie awake worrying about it. Get about 30 minutes of aerobic exercise a day. Drink alcohol in moderation or not at all. Eat foods high in antioxidants (green leafy vegetables, dark-colored fruits like blueberries). Limit intake of foods high in fats, sugars, salt, and refined flours. Do not smoke and avoid people who do smoke, at least while they are smoking. Maintain a healthy body weight—not too thin, not too fat. Hang out with people you like, especially well-behaved young children. Find times to meditate, do yoga, or just relax and be quiet for a few minutes every day. Do not consume too many stimulants like coffee and caffeinated drinks. Floss. Immerse yourself in "nature" several times a week. Wear sunscreen and seatbelts. Find time to be playful. Bet you did not need to read this book to know these recommendations. Not much is new—just about everything that will enhance health at any age will also enhance longevity.

11

Implications for Women's Health in the 21st Century—and Preventing the Epidemiological Collision

I have discussed a number of sobering concepts and ideas throughout this book. Included among them is the incontrovertible evidence that developmental events and the biological and sociocultural factors that affect them have not only lifelong effects on health and reproduction but also transcend generations to influence the health and reproductive success of grandchildren and great-grandchildren. Does that mean that we should just throw up our hands and quit trying to make improvements in health for ourselves and our children? I suppose a pessimistic answer would be yes, especially if we come away from this discussion with an idea that our biology is fixed in stone and that we cannot do anything about it because we can do nothing about the past. But as has often been said, knowing something about our DNA does not tell us our future; rather, it tells us about our past.[1] Knowing something about the conditions under which our reproductive systems developed and evolved tells us something about our past, but does not necessarily constrain our futures. If you remember nothing else about an evolutionary approach to women's health, remember that natural selection has favored adaptability and flexibility, and as such we have a lot of control over our health and the health of our descendents. Perhaps somewhere between our evolved bodies and our 21st-century bodies is a sort of "happy medium" that represents a "good enough" way of living. Where on that arc a given person finds a place to light is variable and depends on individual as well as sociocultural factors. But some of the information in this book may help in finding that resting spot.

As I have asserted repeatedly, in the evolutionary view of life, there can never be perfection. Most of us live lives that are "good enough." We cannot be perfect mothers because something else will have to be compromised too much. We cannot reach reproductive maturity at the very best time, but at a "pretty good" time, all other things considered. The way each of us approaches or experiences health, development, and reproduction are all the results of trade-offs of time, energy, and resources. We think this way all the time: I cannot be both a great anthropologist and a good mother, at least not at the same time, so something has to give. I cannot have a job I love and still be able to spend several months a year traveling. Juggling time and energy is something we do all day long and we accept it as part of modern life. But this juggling act is not just modern and Western; it has been going on forever and describes every living organism from the single-celled amoeba to humans.

We have seen that a woman's biology is the result of the evolutionary history of the human species, the environment and resource conditions under which she grew up and in which her reproductive system developed, and factors in her ongoing daily life, her "lived experience." Because of evolutionary forces, her biology has been shaped to increase her reproductive success and to maintain the ability to reproduce under varying conditions. Women who stop menstruating when they are training for marathons have reproductive systems that are working exactly the way they should be working because reproductive suppression is advantageous under extreme energetic stress, or at least it was in the past. Clinically, such cases were often treated with hormones "to get the system working again." This may have placed some women at additional risk. This is one of the ways in which under-standing our evolved biology can lead to better health care options in treating such "disorders" as infertility (infecundity), many of which are actually healthful responses. As anthropologist Virginia Vitzthum summarizes it, "Taken together, the inevitable, and somewhat daunting, conclusion is that we can expect to find considerable variation in human reproduction. This variation is neither patholo-gical nor paradoxical. Rather, it is an extraordinary adaptation to the varying world in which our species evolved."[2]

Too Much of a Good Thing? Reproductive Hormones and Health

Many of the things we are concerned about with regard to women's health (at least in health-rich nations) may come down to levels of ovarian hormones to which women are exposed throughout their lives. For women who are contracepting and who have only one or two children whom they breastfeed for three to four months, or who have no children at all, the cycling condition characterizes the majority of their reproductive lives, with only a year or two of the hormonal profile character-ized by pregnancy and lactation (see Figure 2-2 in Chapter 2). Thus, they have very little relief from the ups and downs of ovarian hormones (especially estrogen and progesterone) and the associated breast and uterine cell turnover. Women throughout history and in many populations today spend most of their lives in the hormonal milieu that characterizes pregnancy and lactation, with only a few months of exposure to the cycling condition. Furthermore, as I have noted, not only are there more rapid changes in hormone levels for contracepting women in health-rich populations, but the absolute levels are higher throughout all conditions.

Activity levels, developmental history, reproductive history, and diet have a lot to do with the levels of reproductive hormones to which we are exposed (Figure 11-1). This begins while we are in utero with exposure to our mothers' high levels of hormones. (Actually, as noted, it begins with her embryonic exposure, which begins with her mother's embryonic exposure and so on back into the depths of time.) When age of menarche is early, age of menopause is late, age at first birth is late, birth intervals are long, and months of breastfeeding are few, lifetime exposure to ovarian hormones is high. These levels seem to be further elevated when energy intake is high and energy expenditure is low, resulting in lifetime positive energy balance.

Vegetarian diets, particularly when fiber intake is high, are associated with lower levels of estrogen.[3] Low overall caloric intake is also associated with lower estrogen. In contrast, high-fat diets are associated with higher levels. These dietary patterns are related to evolved reproductive strategy in that they signal energy scarcity (vegetarian, high fiber, restricted calories) or plenty (high fat). Fiber, in particular, seems to have an effect on estrogen levels, perhaps due to the phytoestrogens that are common constituents, some of which serve to lower estrogen levels.[4] Quantity of fiber seems to matter. At low levels, phytoestrogens stimulate estrogen production but at high levels, they inhibit it.[5] Additionally, fiber and phytoestrogens consumed by a woman during pregnancy can have the effect of increasing or decreasing cancer risks in her daughters. Fiber intake is notoriously low in many Western diets.

Independent lines of evidence point to a link between ovarian hormones and cancer risk, including the protective effect that seems to occur when the ovaries are removed. Risk for reproductive cancers is reduced following menopause but exogenous hormones, such as those used in HRT and high-estrogen contraceptives, may increase risk. Obesity is linked to high levels of ovarian hormones and reproductive cancers. Cancer growth is more vigorous in younger women who have higher hormone levels than older women in the same population. Cancer risk is higher in pregnancy (when hormone production is high) and lower in lactation (when steroid hormones are suppressed). Further, some of the most successful treatments for reproductive cancer rely on methods of suppressing endogenous hormones.[6]

Medical research and practice has only recently begun to appreciate the role that infectious agents play in various cancers. For example, human papillomavirus (HPV) has been linked to squamous cell skin cancer and infectious processes are implicated in cancers of the stomach, liver, cervix, mouth, and nose.[7] Paul Ewald suggests that the high levels of progesterone during cycling and pregnancy suppress immune function, exposing women to greater risks of infection, especially with Epstein Barr Virus (EBV) and HPV, both of which have been linked to breast cancer. Infections with these viruses can inhibit barriers to carcinogenesis.

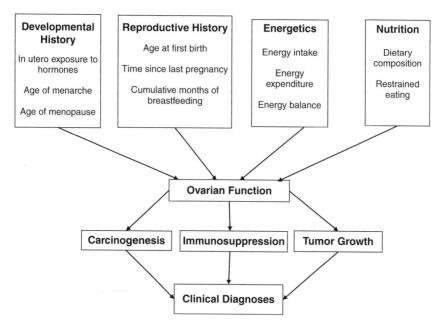

Figure 11-1 Reproductive ecology and ovarian function. Modified from Ellison, 1999, page 203.

Figure 11-2 summarizes some of the health challenges that seem to be associated with[8] chronic exposure to high levels of reproductive hormones. These high levels characterize women in health-rich populations that have been experiencing adequate nutrition, low to moderate physical activity, and good health care with few infections for generations—in other words, most of the women reading this book. There is not much we can do about the hormone exposure we experienced in utero, but we can do something with those arrows that connect hormones to reproductive and health phenomena. This is where the amazing plasticity and resilience of the human species comes in. We have seen that lowering fat intake and increasing physical activity (exercise) has the effect of lowering hormone levels. This is, of course, a recommendation that would be proffered for just about every chronic ailment for which women in health-rich populations are at risk.

Following the arrows in the figure, which represent associations rather than causation, we have seen that high levels of ovarian hormones have been implicated in the longer flows and higher rates of menstrual-cycle associated problems such as cramps and PMS. In the latter case, it may be that PMS is more noticeable in women who experience steeper "falls" in progesterone levels at the end of a cycle. For women with high levels of progesterone in pregnancy, the fall that accompanies the end of gestation may yield higher rates and intensity of postpartum low mood. At menopause the cessation of production of high levels of progesterone and

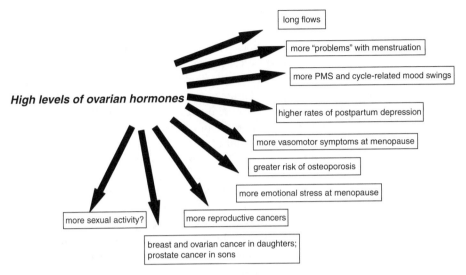

Figure 11-2 Health challenges that seem to be related to high levels of ovarian hormones in health-rich populations. Arrows represent presumed associations and do not necessarily indicate causation. For example, reproductive cancers like breast cancer are linked to genes, diet, cell turnover rates, and dozens of other factors in addition to high hormone levels. It would be inappropriate at this stage of our knowledge to say that high hormone levels "cause" cancer.

estrogen (another steep drop) may be associated with higher reported frequency of vasomotor problems (hot flashes and night sweats) and even low mood that seem to occur in women from health-rich populations. Developing fetuses that are exposed to high maternal hormones seem to be at greater risk for breast and ovarian cancers in girls and prostate cancer in boys. Finally, high levels of estrogen can speculatively be related to the reported (although not confirmed) higher levels of sexual activity in health-rich populations. All of these associations are points that can be explored with further research, for example, with athletes and nonathletes from a single health-rich population.

Can We Change Our Behaviors to Improve Our Health?

On one level, making the lifestyle changes that are more consonant with our ancestral lives and will lead to improvements in health is up to individual effort. But we know full well that this "blaming the victim" approach is not the whole story and, in fact, if there are not community and institutional changes to accompany individual efforts, improved population health will be slow in coming. One of the most formidable challenges to improving women's health worldwide is confronting the complex political and ideological realities that influence the level of control that women have over their lives. This book has given short shrift to these all-important factors, but they cannot be ignored if we have any hope of improving health. For example, we understand from world surveys that the single most successful way of improving lives is the education of women and girls. But the desired systemic changes are slow in coming and our children and grandchildren do not have time to wait. Ultimately, it comes down to individual decisions, no matter how difficult the challenges. Many women can make major changes when they are pregnant: they quit smoking and drinking, try to eat more healthfully, and get out and walk around the block a few times every week. They can do this because there is someone else whose life and health depends on them. From the old aphorism "the life you save may be your own" (apologies to Flannery O'Connor) we now have "the lives you save may be those of your children, grandchildren, and great grandchildren."

Actually, people in health-rich populations are pretty good at taking care of themselves when they really want to. Smoking rates have gone down impressively since the middle of the last century—down by half from the mid-1960s to the mid-1990s in the United States (52% in 1965 to 26% in 1999 for men over 18 in the United States). Exercise is increasing slightly in this country. Drinking while intoxicated (DWI) rates seem to be slowly decreasing in my home state of New Mexico (which formerly ranked #1 in the United States in drunk driving) and there has been a decline in alcohol-related fatalities in recent years. The numbers of pregnant women who seek prenatal care each year is on the rise and women are seeking this care earlier in their pregnancies. In less than 100 years, drinking water and sanitation improved, immunization rates went up, and infant mortality went down. These are all signs that things are getting better, in contrast to the trends described earlier toward increasing rates of obesity, type 2 diabetes, and reproductive cancers. Rather than focus all of our research dollars and efforts on "curing" these diseases and disorders, we should channel some toward prevention. It is a mantra that haunts every field of medicine, but one day we may heed it and find that our efforts pay off.

The truth is, we do not know what really makes people change their behaviors, despite decades of research in health psychology and behavioral health.[9] I cannot tell you what that dentist in Gunnison, Colorado, said to me in 1978 that led me to floss my teeth every single day of my life since then (more than 30 years!), but it had an impact on me. I'm guessing that he said pretty much the same thing that my parents and other dentists had said to me for years, but for some unknown reason, I was receptive that one time and it had a lifelong impact. A student of mine once told me that an offhand comment that I made about her smoking (probably mildly derogatory) had an impact on her that nothing anyone had ever said before had—she quit smoking immediately. It probably was not really anything that I said, but for one reason or another, she was susceptible to hearing something that would lead her to make a major behavioral change. (I continue to make mildly derogatory comments about smoking in hope that the scenario will repeat itself, but so far, I do not have any evidence that it has made a difference—except to annoy my friends and students.) So if only a few of the readers of this book sift through the information to find bits and pieces that are relevant to their lives and make changes that improve their own health and that of their children and grandchildren, then all of the excellent research that my friends and colleagues have done on women's health over the years will have paid off.

Studies of nonhuman primates suggest that the life stage at which an individual is most likely to adopt new behaviors is the juvenile stage (adolescence or teenage in humans). It is also the age-group that in humans is most susceptible to behaviors that may not be positive for later health (smoking, alcohol consumption, drug use, "junk food" consumption, extreme dieting, truancy, dropping out of school). Social epidemiologists Mary Schooling and Diana Kuh have developed a model of the pathways between childhood and women's health behaviors that suggest that the best (or at least a very good) place to target interventions might be adolescence.[10] They coin the phrase *behavioral capital* to describe the positive behaviors and attributes that can be acquired during adolescence and that seem to have lifelong and transgenerational positive effects on health. These include "social competence, decision-making and problem-solving skills, coping strategies, personal efficacy, self-esteem, attitudes and values that help the individual remain resilient in times of adversity or take advantage of talents and opportunities."[11] These behaviors lead not only to greater likelihood of making positive lifestyle choices but also to greater educational and employment opportunities that are themselves related to positive health. Adults with these attributes often make positive decisions about parenting and pass on to their children the behavioral capital that they themselves have. Like financial capital, behavioral capital can be inherited and passed along to future generations. Consider that a teenage girl who "decides" to not smoke, eat a healthful diet, and delay pregnancy until she is in her early 20s is behaving in a way that could affect the health of her descendents who could number in the thousands in just a few generations.

To some it might seem acceptable to provide the information and, to the extent possible, the social and medical support to enable people to make healthful life changes, but if they continue to engage in risky behaviors, that is their "right," so let them continue to live the lives they choose. But now that we know about the long-term and transgenerational consequences of lifestyle choices or impositions, should we simply stop once we have provided the information?

An evolutionary perspective tends to ignore some of the most important factors that impact women's health, which include socioeconomic status, cultural norms, education, media and peer influences, and geopolitical forces. On the other hand, most sociocultural and epidemiological models of health ignore evolutionary and transgenerational factors. I have modified the figure from Schooling and Kuh to add what we know about evolutionary history to the flowcharts that have been developed to show influences on health. In Figure 11-3, for example, the far left (first) box on the original chart from Schooling and Kuh was labeled "childhood

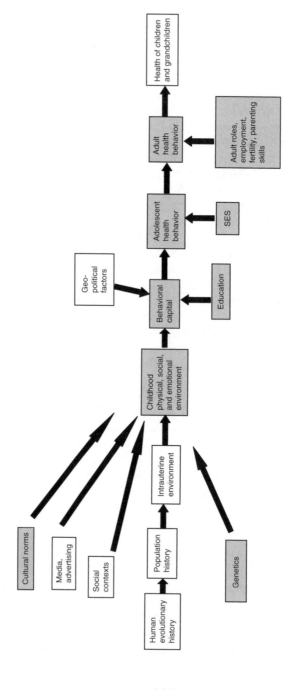

Figure 11-3 Evolutionary and selected individual lived experience pathways to lifelong and transgenerational health for women. Adapted and modified from Schooling and Kuh, 2002, page 284 (shaded boxes are in the original).

environment." For much of what we do in trying to improve individual and population health, this is certainly a good place to start. But each of us is born with an evolutionary history that has its own impacts, by providing both constraints and potentials. We cannot really change that evolutionary history, but we can modify behaviors and practices to improve what we are stuck with so that when the cycle repeats itself every generation, the overall healthfulness is ratcheted up a bit each time.

Woman the Baby-Maker

Some readers may complain that in this book I have replaced the "Man the Hunter" model of human evolution with a "Woman the Baby-Maker" model. If you believed that, you may not have reached this point in the book. But what I hope I have conveyed is that although women are shaped and, in some cases, constrained by the way natural selection has operated on their bodies to increase reproductive success, this does not mean that our only lot in life is reproducing; it simply tells us where to look in our evolutionary history for characteristics that affect our health today in environments of plenty. Like DNA and intrauterine gestation, knowing about our evolutionary history does not tell us about our future; it tells us about our past. We may be "stuck with" a pelvis better suited to bipedal walking than to giving birth to large-brained babies, but we can deal with that by providing emotional and technological support, even at the level of delivering babies surgically and bypassing the restrictive birth canal.

But we are more likely to develop ways of promoting positive health if we know and understand genetics, gestational environments, and our evolved biology. We cannot deny our legacies from our "inner fish"[12] all the way to our "inner Paleolithic *Homo*." But we are not helpless victims of these inner whatevers; rather, we have amazing minds, an incredible degree of biological flexibility, and the all-important phenomenon of culture that we can use to transcend the unfortunate legacies of our evolutionary history and embrace the good ones. We also have individual autonomy, whether a great deal or only a tiny bit, that we can use to improve our lives and those of our families and our communities. For those of us who have benefited from the incredible opportunities offered us by being born in health-rich nations, our behaviors can have far-reaching impacts.

The Epidemiological Collision

Changing socioeconomic conditions accompanying the industrial revolution resulted in what has been called an epidemiological transition reflected in changing disease profiles and causes of death from primarily infectious to primarily degenerative. Before this transition, infant mortality was high, fertility was high, and life expectancy was low. Most causes of mortality were infectious and life expectancy was lower for women than for men.[13] After the transition, people began living longer and dying of chronic and degenerative diseases like cardiovascular diseases, cancers, and stroke. Improved standards of living (like diet and hygiene), behavioral modifications, and improved health care all contributed to the changes in health profiles. We also know that much of the change in health and disease profiles is traceable to reproductive factors, including age of menarche and first pregnancy, and diet during pregnancy. Because these underlying factors vary from population to population, different parts of the world have passed through the epidemiological transition at different rates: Western Europe and the United States led the way, followed by southern and Eastern Europe and parts of Asia. Some populations have not yet made the transition and have health profiles characterized by high rates of infectious diseases, high fertility, high infant mortality, and low life expectancy, associated with poor nutrition and limited access to health care.

The history of the epidemiological transition leads us to expect the rest of the world to "catch up" with the health-rich nations by passing through the same kind of transition that they faced a hundred or so years ago. What we are seeing instead might be more appropriately called an "epidemiological collision" in nations like Zimbabwe, Haiti, and Ecuador, where infectious diseases are still a problem, childhood malnutrition is rampant, and obesity is on the increase. Associated with increased obesity, of course, is the expected increase in hypertension, type 2 diabetes, cardiovascular diseases, and cancers. In other words, many nations have the worst of both worlds—the poverty, pathogens, and malnutrition of the health-poor world but also the chronic and degenerative diseases of the health-rich world. Table 11-1 provides an overview of health-related data from 19 nations to illustrate this potential collision. The table shows that type 2 diabetes, usually seen as a problem for health-rich populations, is likely to increase in otherwise health-poor nations as obesity rates increase. Note, for example, Zimbabwe and Pakistan, where infectious diseases surpass chronic diseases as causes of death and type 2 diabetes is not among the top 10 killers. But note also that obesity is on the rise, especially in Zimbabwe. Sadly, perhaps the only reason that type 2 diabetes is not in the top 10 in that country is that average life expectancy is only 34.

Projections for world health in 2030 tend not to be very optimistic. Most assume that obesity will continue to increase along with globalization. As women in health-rich nations have decreased levels of physical activity over the last few decades, so too will women in populations in transition. This means that they will begin to suffer from the "diseases of civilization" that characterize health-rich areas of the world. As populations aspire to achieve the standard of living and life expectancy of health-rich populations, they will also see similarities in rates of type 2 diabetes, obesity, hypertension, and cancers. As an example, when I first went to the Philippines in the 1970s, I could find no one who had ever heard of type 2 diabetes. It was rare when I returned in 1987 and was beginning to increase when I visited again in 1999. In 2000, there were almost 3 million Filipinos with type 2

Table 11-1 Selected health data for 19 nations

Country	Rich:poor in top 10 causes of death	BMI ≥ 25 and ≥ 30 kg/m^2 (women 14–49)		Type 2 diabetes in top 10? (2002)	Under 5 mortality (per 1,000)	Maternal mortality (per 100,000)	Life expectancy (females)
Afghanistan	3:5	17.4	1.8	no	256	1,900	42
Australia	9:1	62.7	24.9	#7	5	6	83
Bangladesh	3:5	5.4	0.2	no	73	380	63
Cambodia	3:7	9.3	0.1	no	127	450	58
China	6:3	24.7	1.8	no	36	56	74
Cuba	7:1	61.1	24.6	no	7	33	80
Ecuador	6:3	52.6	16.8	#3	24	130	75
Haiti	3:6	50.6	15.0	#10	112	680	56
India	3:5	15.2	1.4	no	89	540	63
Japan	8:1	18.1	1.5	no	3	10	86
Mauritius	7:2	52.3	18.3	#3	14	24	75
Mexico	6:2	67.9	34.3	#1	25	83	77
Pakistan	3:6	25.5	3.6	no	100	500	63
Philippines	5:4	28.5	3.7	#9	28	200	72
Saudi Arabia	5:3	63.8	33.8	#6	24	23	74
South Africa	4:4	67.2	35.2	#9	62	230	49
USA	8:1	72.6	41.8	#6	7	14	80
Zambia	2:7	18.6	1.3	no	750	173	40
Zimbabwe	2:7	48.8	15.3	no	121	1,100	34

"Rich" diseases include cardiovascular diseases, type 2 diabetes, cancers, stroke, and Alzheimer's.
"Poor" diseases include respiratory illnesses, perinatal conditions, diarrhea, tuberculosis, and malaria.
Traffic accidents, self-inflicted deaths, war, violence, and congenital diseases aren't included in this ratio when they are in the top 10.
From World Health Organization statistical database.

diabetes and that number is projected to reach almost 8 million by 2030.[14] Much of that increase is due to increases in obesity and certain to be associated with that are higher levels of ovarian hormones. Women in the Philippines are likely to have high birth rates, however (it is a Catholic country), and so lifelong exposure may not be great enough to lead to breast cancer rates like those seen in health-rich nations. Equally important, however, is maintaining high rates and duration of breast-feeding as a way of further minimizing lifelong exposure to hormones. Unfortunately, UNICEF reports that only 16% of Filipinas were breastfeeding their children at 4–5 months of age in 2003.

In the United States we are struggling with rapidly rising costs and gross inequalities in health care for Americans. Among the reasons for this current "health crisis" is increasing burden from chronic diseases like those discussed in this book. If anything, the burden is going to increase as the population ages. Furthermore, the causes of death and ill health in the rest of the world will look increasingly like those in the United States, and health care systems in most health-poor nations are already struggling and far from adequate. Ethical issues also arise as medical technology becomes ever more sophisticated. Do we give the liver transplant to the 85-year-old man or the young mother of three? No one has an easy answer to this question and it is an ongoing struggle for medical ethicists. How do we value one life above another, no matter what the circumstance? Do we put our money into worldwide immunization programs or efforts to reduce cardiovascular disease? Is "good enough"[15] health good enough?

Many scholars who investigate the relationships among various lifestyles, environmental, political, and health factors argue that there are insufficient data to make policy or practice recommendations. Furthermore, those of us who view health through the lens of evolution do not see ourselves "practicing" evolutionary medicine by making prescriptive suggestions,[16] but we have now seen how much effect early life experiences have on risk for various late-life diseases and disorders, suggesting that future research and efforts toward prevention should focus on these stages of the life course.[17] This means more attention to pregnancy as a way to reduce the risk for female fetuses; continued effort to increase rates and duration of breastfeeding; and providing adequate nutrition during childhood. As with many recommendations that come from an evolutionary approach to women's health, efforts to improve prenatal, infant, and childhood health have almost no known negative consequences, but they do have far-reaching benefits. Along with reduced risk for reproductive cancers would come reduced risks for cardiovascular disease, infertility, type 2 diabetes, hypertension, and other cancers. If we invested trillions of dollars and all the intellectual talent on earth, we would be highly unlikely to develop a magic bullet that would have equivalent effects.

Isn't it time we paid attention to what our evolved bodies have been telling us?

Notes

Introduction

1. Gluckman and Hanson, 2006; Eaton, Shostak, and Konner, 1988.
2. Baker, 2008.
3. Gluckman and Hanson, 2006.
4. Barrett et al., 1998.
5. Knott, 2001.
6. Williams and Nesse, 1991; Nesse and Williams, 1994; 1997.
7. See reviews in Trevathan, Smith, and McKenna *Evolutionary Medicine and Health: New Perspectives*, 2008, and Trevathan, 2007.
8. Nesse, 2001; Nesse and Schiffman, 2003; Nesse, Stearns, and Omenn, 2006.
9. Nesse and Stearns, 2008.
10. Nesse and Stearns, 2008, page 30.
11. Nesse and Stearns, 2008, page 28.
12. Stearns and Ebert, 2001.
13. Ewald, 1994.
14. Ewald, 1999.
15. Zuk, 1997.
16. Martin, 1987.
17. Trevathan, 2007.
18. Kluger, 1978.
19. Williams and Nesse, 1991.
20. For a particular gene, each of us has one version from our mother and the same or a different one from our father. Different versions of genes are called "alleles" and one from each parent combine at conception to result in a gene for a specific trait. If a gene has two of the same alleles, it is homozygous for a trait; if it has two different alleles, it is heterozygous.
21. Livingstone, 1958.
22. Ewald, in press.
23. Ewald, 2008.
24. Stearns, 1992.
25. Mace, 2000.
26. Daly and Wilson, 1983; Lack, Gibb, and Owen, 1957.
27. Walker et al., 2008.
28. Walker et al., 2008, page 831.
29. Valeggia and Ellison, 2003.
30. Bentley, 1999.

31. Sievert, 2006.
32. Robson, van Schaik, and Hawkes, 2006.
33. Bentley, 1999, page 183.
34. Jurmain et al., 2008.
35. Olshansky, Carnes, and Butler, 2001.
36. Ackerman, 2006, citing Craig Stanford.
37. Barash, 2005.
38. Krogman, 1951.
39. Van Schaik et al., 2006; Allman and Hasenstaub, 1999.
40. Robson, van Schaik, and Hawkes, 2006.
41. Crews, 2003.
42. Hrdy, 2005.
43. Hrdy, 2005.
44. Hrdy, 2005; Hawkes, 1991.
45. Cohen, 2003.
46. Hrdy, 2005.
47. Mace, 2000.
48. Finch and Rose, 1995.
49. Estrogens are a class of steroid hormones that include estradiol, estrone, and estriol. For simplification I will use the general term *estrogen* to refer to all members of the class throughout this book.
50. Ellison, 1988.
51. Worthman and Stallings, 1994; 1997.
52. I first encountered the terms in Caleb Finch's book *The Biology of Human Longevity* and he communicated that he "devised it in frustration Oct 2006" (personal communication, February 9, 2009).

Chapter 1

1. Thomas et al., 2001.
2. Bogin and Smith, 1996.
3. Golub, 2000.
4. Worthman, 1998.
5. Bogin et al., 2002.
6. Bogin and Loucky, 1997.
7. Diamond, 1992.
8. I saw this myself in three successive trips to Japan. On my first trip, in 1971, at 5'7" I towered above almost everyone in the train stations. On my second trip in 1980, there were a lot of people my height, but we were still in a minority. On a third trip in 1990, it felt no different from being in a train station in the United States. Improved health and nutritional status since World War II has been cited as the reason that Japanese stature has converged on that of the West.
9. Bogin, Varela-Silva, and Rios, 2007.
10. Bogin and Loucky, 1997.
11. Worthman, 1999.

12. Frisch and Revelle, 1970.
13. Frisch and McArthur, 1974.
14. Ellison, 2001.
15. Dufour and Sauther, 2002.
16. Bronson, 2000.
17. Dufour and Sauther, 2002.
18. Bronson, 2000.
19. Lassek and Gaulin, 2007.
20. Lassek and Gaulin, 2008.
21. Lassek and Gaulin, 2008, page 26.
22. Brown and Konner, 1987.
23. Burger and Gochfeld, 1985.
24. Chisholm and Coall, 2008.
25. Whether or not humans are affected by pheromones is a hotly debated topic as will be discussed in the next chapter with regard to the phenomenon of menstrual synchrony.
26. Surbey, 1990.
27. Draper and Harpending, 1982.
28. Surbey, 1990.
29. Belsky, Steinberg, and Draper, 1991; Moffitt et al., 1992.
30. Borgerhoff Mulder, 1989.
31. Apter and Vihko, 1983.
32. Worthman, 1999.
33. Liestøl, 1980.
34. Wyshak, 1983.
35. Gardner, 1983.
36. Ellis, 2004.
37. Ellis, 2004.
38. Vihko and Apter, 1984.
39. Berkey et al., 2000.
40. Wiley, 2008.
41. Law, 2000.
42. Barker et al., 2008a.
43. Barker et al., 2008b.
44. Kuzawa, 2007; 2008.
45. Graber et al., 1997.
46. Graber et al., 1997 for the United States; Kaltiala-Heino et al., 2003 for Finland; Wichstrom, 2000 for Norway.
47. Angold, Costello, and Worthman, 1998.
48. Worthman, 1999, page 152.
49. I should also note that the answer to the question of when puberty should occur is very different from the perspective of modern medicine, which focuses on health, and the perspective of evolution, which focuses on reproductive success.
50. Eaton et al., 1994.
51. Stearns and Koella, 1986.
52. Moerman, 1982.
53. Scholl et al., 1994; Friede et al., 1987.

54. Wallace et al., 2004.
55. Kramer, 2008, page 339.
56. Geronimus, 1987; King, 2003.
57. Phipps and Sowers, 2002; Geronimus, 2003.
58. Kramer, 2008.
59. Kramer, 2008.
60. Geronimus, 2003; Wiley and Allen, 2009.
61. Barber, 2001.
62. Barber, 2002.
63. Geronimus, 2003.
64. Geronimus, 2003.
65. Jordan, 1997, page 58, also cited in Geronimus, 2003.
66. Hrdy, 1999.
67. Ellison, 1999.
68. Coall and Chisholm, 2003.
69. Armelagos, Brown, and Turner, 2005; Barrett et al., 1998.
70. Chisholm and Coall, 2008.
71. Worthman, 1999.
72. Chisholm and Coall, 2008; Chisholm, 1993.
73. Worthman, 1999, page 154.

Chapter 2

1. Strassmann, 1999a.
2. Short, 1976, page 3.
3. Shostak, 1981.
4. Short, 1976.
5. Eaton et al., 1994; Eaton and Eaton, 1999; Ellison, 1999.
6. Eaton et al., 1994; Eaton and Eaton, 1999; Ellison, 1999.
7. Eaton et al., 1994.
8. Eaton and Eaton, 1999.
9. Vitzthum, 2008.
10. Vitzthum, Spielvogel and Thornburg, 2004.
11. Standard clinical practice in the United States considers progesterone levels <40% of "normal" levels to be indicative of an anovulatory cycle. For example, based on a study of healthy Chicago women, cycles <132 pmol/l peak progesterone level should be classified as anovulatory. Yet in the Bolivian high-altitude sample studied by Vitzthum and her colleagues (and using the same assay protocols as those in the Chicago study), normal ovulatory cycles were observed to have peak progesterone levels as low as 111 pmol/l.
12. Jasienska and Thune, 2001.
13. Knott, 2001.
14. Vitzthum, 2001; Vitzthum and Spielvogel, 2003.
15. Bribiescas and Ellison, 2008.
16. Vitzthum, 2008.

17. Gibson and Mace, 2005.
18. Kramer and McMillan, 2006.
19. Whitten and Naftolin, 1998.
20. Thompson et al., 2008.
21. Thompson et al., 2008.
22. Vitzthum, 2008; Ellison, 1990.
23. Wasser and Isenberg, 1986.
24. Vitzthum, 2008, page 66.
25. Núñnez-de la Mora and Bentley, 2008.
26. Stearns and Koella, 1986.
27. Vitzthum, 2008; Ellison, 1990. Even anovulatory cycles can sometimes be seen as adaptive for temporarily stressful times such as occur with resource seasonality as pointed out by Vitzthum et al., 2002. In fact, as noted in Chapter 2 in our discussion of body fat, patterns of birth seasonality reported for a number of cultures reflect conceptions that occur during times of abundance (e.g., the harvest or rainy season) when ovarian function may be most robust (Ellison, 2001).
28. Ellison, 1990.
29. Jasienska and Thune, 2001.
30. Vitzthum et al., 2000.
31. Bentley, 1994.
32. Lipson and Ellison, 1992.
33. Pollard, 2008.
34. Ross et al., 1995.
35. Bribiescas and Ellison, 2008; Muehlenbein and Bribiescas, 2005; Campbell, Lukas, and Campbell, 2001.
36. An exception is a book by Coutinho and Segal, 1999 entitled *Is Menstruation Obsolete?*
37. Sievert, 2008.
38. Sanabria, 2009.
39. Sievert, 2008.
40. Buckley and Gottlieb, 1988.
41. Sievert, 2008.
42. Short, 1976.
43. Eaton and Eaton, 1999.
44. Frisch et al., 1985; Wyshak and Frisch, 2000.
45. Jasienska et al., 2006.
46. Martin, 2007.
47. Profet, 1993.
48. Strassmann, 1996.
49. Strassmann, 1996.
50. For a contrasting view, see Clancy, Nenko, and Jasienska, 2006.
51. Finn, 1998.
52. Barnett and Abbott, 2003.
53. Wood, 1994, page 22.
54. Emily Martin (1987) has an interesting reflection on how this process of oocyte loss is described in the medical literature. Most of the terms used to describe the development and decline of the egg are negative: deterioration, breakdown, aging, failure,

degeneration. She points out that, on the other hand, male spermatogenesis is associated with terms like remarkable and amazing. Furthermore, the process associated with ovulation is described as wasteful, whereas the production of millions of sperm per ejaculate with only one destined to unite with an egg impresses with "sheer magnitude." Her analysis is a reminder that even "scientific" descriptions of reproduction are imbued with metaphors that reflect our societal values and attitudes.

55. Wood, 1994.
56. Maynard Smith et al., 1999.
57. Vitzthum et al., 2000.
58. Vitzthum et al., 2001.
59. Martin, 1988.
60. Taylor, 2006.
61. Martin, 1988; Taylor, 2006; Golub, 1985.
62. Taylor, 2006.
63. Taylor, 2006.
64. Reiber, 2008.
65. White et al., 1998.
66. Garland et al., 2003.
67. Ensom, 2000; Doyle, Swain Ewald, and Ewald, 2008.
68. Doyle et al., 2008.
69. Gomez et al., 1993.
70. Trzonkowski et al., 2001.
71. Angstwurm, Gartner, and Ziegler-Heitbrock, 1997.
72. Smith, Baskin, and Marx, 2000; Li and Short, 2002.
73. Kaushic et al., 2000.
74. Fidel, Cutright, and Steele, 2000.
75. Brown et al., 1997.
76. Doyle et al., 2008.
77. Ewald, in press.
78. McClintock, 1971.
79. I conducted an informal survey of undergraduate women students at four universities asking the question "have you ever experienced menstrual synchrony?" Of 280 respondents, 232 (83%) said yes.
80. Goldman and Schneider, 1987; Graham and McGrew, 1980; Quadagno et al., 1981; Weller et al., 1999.
81. Jarett, 1984; Wilson, Hildebrandt Kiefhaber, and Gravel, 1991; Strassmann, 1997; Yang and Schank, 2006; Ziomkiewicz, 2006.
82. Yang and Schank, 2006; Wilson, 1992; Schank, 2000, 2006.
83. Schank, 2006; Strassmann, 1999b; Hays, 2003.
84. Russell, Switz, and Thompson, 1980; Preti, Cutler, and Garcia, 1986.
85. Wilson, 1992.
86. Trevathan, Burleson, and Gregory, 1993.
87. Trevathan et al., 1993.
88. Fisher and others have contributed to a cottage industry of explanations for why human females are "continuously sexually receptive" and "lack estrus," but most of them are not relevant to the discussion at hand.

89. Fisher, 1983.

90. Brewis and Meyer, 2005.

91. Hill, 1988; Pawlowski, 1999.

92. Harvey, 1987; Matteo and Rissman, 1984.

93. Burleson, Trevathan, and Gregory, 2002.

94. Thompson, 2005.

95. Lipson, 2001, page 244.

96. Vitzthum, 2009.

Chapter 3

1. Daly and Wilson, 1983.

2. Gleicher and Barad, 2006.

3. Gleicher and Barad, 2006, page 588.

4. Pollard and Unwin, 2008.

5. Barnett and Abbott, 2003.

6. Nikolaou and Gilling-Smith, 2004.

7. Uvnas-Moberg, 1989.

8. Vitzthum, 2009.

9. Actually, the downside of a dampened immune system may not have been much of a problem until women were having 12–13 menstrual cycles a year with the associated 12–13 times of greater vulnerability to diseases and other infections.

10. Haig, 1993, 1999.

11. Vitzthum, 2009, page 126.

12. Wasser and Isenberg, 1986.

13. Most molecules pass from the mother to the fetus via passive diffusion, but some nutrients are needed in such quantities that diffusion is not sufficient and specific nutrient carrier mechanisms are required, especially later in pregnancy when the fetus is growing rapidly. Nutrients that require these mechanisms include glucose, amino acids, and fatty acids. There is some hope that understanding these nutrient transfer mechanisms may enable the development of ways of treating developmental abnormalities in the fetus, including intrauterine growth retardation (Knipp, Audus, and Soares, 1999).

14. The type of placenta possessed by members of the primate order is one of the characteristics used to divide the order into two suborders, Strepsirhini (*epitheliochorial* placentas with several membrane layers) and Haplorhini (*hemochorial* placentas with thin barriers) (Luckett, 1974).

15. Knox and Baker, 2008.

16. Knox and Baker, 2008.

17. Anderson, 1971.

18. Loisel, Alberts, and Ober, 2008.

19. Loisel et al., 2008.

20. Ober et al., 1992.

21. Wasser and Isenberg, 1986.

22. Type O secretes antibodies against A and B antigens; types A and B secrete antigens against each other.

23. Hanlon-Lundberg and Kirby, 2000.
24. Waterhouse and Hogben, 1947.
25. Chung and Morton, 1961; Matsunaga, 1959.
26. Butler, 1977.
27. Haig, 1993, 1999.
28. Power and Tardif, 2005.
29. Furthermore, it is estimated that only about 12 of every 100 conceptions grow up to reproduce offspring themselves.
30. Vitzthum, 2009.
31. Guerneri et al., 1987.
32. Vitzthum, 2009.
33. Peacock, 1990.
34. Vitzthum, personal communication, February, 2009.
35. Vitzthum, Spielvogel, and Thornburg, 2006.
36. Wasser and Isenberg, 1986.
37. Nepomnaschy et al., 2006.
38. Vitzthum, 2009.
39. Wasser and Isenberg, 1986.
40. Haig, 1999.
41. Tardif et al., 2004.
42. Colhoun and Chaturvedi, 2002.
43. Power and Tardif, 2005.
44. Redman and Sargent, 2005.
45. Robillard et al., 2008; Robillard, Dekker, and Hulsey, 2002.
46. Loisel et al., 2008.
47. Haig, 2008.
48. See for example, Bdolah et al., 2004; Levine et al., 2006; Signore et al., 2006; Venkatesha et al., 2006; Ewald, in press.
49. Robillard et al., 2008.

Chapter 4

1. De Benoist et al., 2004.
2. Harris and Ross, 1987; Whitten, 1999.
3. Profet, 1992.
4. Flaxman and Sherman, 2000.
5. Flaxman and Sherman, 2000.
6. Flaxman and Sherman, 2000; Fessler, 2002.
7. Fessler, 2002.
8. Aiello and Wheeler, 1995.
9. Fessler, 2002, page 26.
10. Flaxman and Sherman, 2000.
11. Flaxman and Sherman, 2000; Furneaux, Langley-Evans and Langley-Evans, 2001.
12. Ewald, in press.
13. Pike, 2000.

14. Forbes, 2002.
15. Flaxman and Sherman, 2008.
16. Wiley and Katz, 1998.
17. Johns and Duquette, 1991.
18. Other food cravings "make sense" from both health and evolutionary perspectives. Cravings for fruits are often reported and fruits serve as good sources of vitamin C. Milk and other diary products are craved in cultures where they are common, providing good sources of calcium.
19. Mills, 2007.
20. O'Rahilly and Muller, 1998.
21. Brandt, 1998, page 164.
22. Kuzawa, 1998.
23. Haig, 2008; Hrdy, 1999.
24. Barker, 1998, 2004.
25. McDade et al., 2001.
26. Adair and Kuzawa, 2001.
27. Seckler, 1982.
28. Martorell, 1989; Wiley and Allen, 2009.
29. Gluckman, Hanson, and Beedle, 2007.
30. Kuzawa, 2005.
31. Ellison, 2005.
32. Nathanielsz, 2001.
33. Finch, 2007.
34. McDade, 2005.
35. Whitcome, Shapiro, and Leiberman, 2007.
36. Seckl, Drake, and Holmes, 2005.
37. McDade, 2005.
38. Davis et al., 2005.
39. It should be noted that most studies of the effects of prenatal stress on later behavior have been done on species other than humans (Weinstock, 2001), and the findings may not all apply to humans.
40. Nathanielesz, 2001.
41. Glynn et al., 2001.
42. Clapp et al., 1992; Wang and Apgar, 1998.
43. Baker, 2008.

Chapter 5

1. Lovejoy, 2005; Schimpf and Tulikangas, 2005.
2. Abitbol, 1987a, b.
3. Stewart, 1984, page 611.
4. Pawlowski and Grabarczyk, 2003.
5. Abitbol, 1987b.
6. Lovejoy, 2005.
7. Walrath, 2003.

8. Sheiner et al., 2004.

9. Trevathan, 1987.

10. Rosenberg and Trevathan, 1996, 2001, 2002.

11. Trevathan and Rosenberg, 2000.

12. An exception may be the Neandertals, who are notable for being very stocky and broad shouldered and for having broad pelvises. It may be that the large pelvic entry (formed by the long pubic bone) is an adaptation to giving birth to broad-shouldered infants (Rosenberg, 1992).

13. Trevathan, 1987, 1999.

14. Trevathan, 1987, 1999; Rosenberg and Trevathan, 1996, 2002.

15. Nesse, 1991.

16. Steer, 2006, page 137.

17. DeSilva and Lesnik, 2008.

18. Stoller, 1995.

19. Lovejoy, 2005.

20. Locke and Bogin, 2006.

21. Bjorkland, 1997.

22. Schimpf and Tulikangas, 2005; Abitbol, 1988.

23. The ischial spines are landmarks used by midwives and obstetricians to mark the progress of the fetus as it descends through the pelvis. When the head is at the ischial spines, it is said to be at zero (0) station. Above is minus 1, 2, or 3 and below, plus 1, 2, or 3. There are three pelvic planes that are identified in this process: the pelvic inlet, midplane, and outlet. The narrowest point is the midplane, delineated by the pubic symphysis in front, the sacrum in back, and the ischial spines on the sides. This is the most likely place for fetal descent to be arrested in labor.

24. Stewart, 1984, page 616.

25. Dietz, 2008.

26. Nygaard et al., 2008.

27. Handa et al., 2003.

28. Although cesarean section has been touted as a way to protect the pelvic floor, there is not good evidence that this is the case. Other technological interventions such as episiotomy and instrumental delivery appear to be far more important in weakening the pelvic floor than cesarean section is for leaving the floor undamaged (Goer, 2001).

29. Abitbol, 1988, page 59.

30. Abitbol, 1988.

31. The only other mammal that seems to be susceptible to pelvic organ prolapse is the squirrel monkey, but the mechanism involved seems to be different from that observed in humans (Williams, 2008). Furthermore, POP has only been reported for captive squirrel monkeys, so it is not clear that it occurs in the wild.

32. Whitcome et al., 2007.

33. Abitbol, 1993.

34. Graafmans et al., 2002.

35. Blurton Jones, 1978.

36. Walsh, 2008.

37. Joseph Walsh has proposed the hypothesis that the "release" from natural selection against large babies afforded by c-section may yield an increase in IQ

in those who are delivered surgically. He notes that newborn brain size corresponds to newborn weight and that there is a direct correlation between birth weight and IQ at 7 years in children who were of normal or higher weight at birth.

38. Read et al., 1994.
39. Liston, 2003.
40. Liston, 2003.
41. Schuitemaker et al., 1997.
42. Hannah, 2004.
43. Varner et al., 1996.
44. Freudigman and Thoman, 1998.
45. Swain et al., 2008.
46. Cardwell et al., 2008.
47. Liston, 2003.
48. Doherty and Eichenwald, 2004.
49. Hannah, 2004.
50. Montagu, 1978.
51. Montagu, 1978.
52. Weiss, 2008; Lagercrantz and Slotkin, 1986.
53. Unfortunately, some of the responses to catecholamines may look like fetal distress. A common reason for cesarean section is evidence of heart-rate changes in the fetus. When electronic fetal monitoring became a standard part of labor and delivery in hospitals, the cesarean section rate increased in the United States. It turned out that in most cases the heart-rate changes were normal responses to catecholamine release and did not indicate life-threatening situations (Lagercrantz and Slotkin, 1986).
54. Lagercrantz and Slotkin, 1986.
55. Lagercrantz and Slotkin, 1986.
56. Varendi, Porter, and Winberg, 2002.
57. Winberg, 2005.
58. Varendi et al., 2002.
59. Lagercrantz and Slotkin, 1986.
60. Marchini et al., 2000.
61. Doherty and Eichenwald, 2004.
62. Roy, 2003.
63. Trevathan, 1987.
64. Trevathan, 1988.
65. Michel et al., 2002.
66. Simkin, 2003.
67. Baxley and Gobbo, 2004.
68. Beer and Folghera, 2002; Beer, 2003.
69. Walrath, 2006.
70. Odent, 2003.
71. Montagu, 1961.
72. Portmann, 1990.
73. Montagu, 1964, page 233.

Chapter 6

1. Martin, 2007.
2. Portmann, 1990. (Portmann first coined the phrase in a 1942 German publication.)
3. Trevathan, 1987.
4. Schultz, 1949.
5. Winberg, 2005.
6. Uvnas-Moberg, 1996.
7. Uvnas-Moberg, 1996.
8. Trevathan, 1987, p. 213, italics in the original.
9. Trevathan, 1981, 1987.
10. Uvnas-Moberg, 1996.
11. Mikiel-Kostyra, Mazur, and Bołtruszko, 2002.
12. Tollin et al., 2005; Marchini et al., 2002.
13. Hoath, Pickens, and Visscher, 2006.
14. Westphal, 2004.
15. Visscher et al., 2005.
16. Hoath et al., 2006.
17. Tollin et al., 2005.
18. Tollin et al., 2005, page 2397.
19. Tansirikongkol et al., 2007.
20. Bystrova, 2009.
21. Zasloff, 2003.
22. Hoath, Narendran and Visscher, 2001.
23. Hoath et al., 2006.
24. World Health Organization, 2006.
25. Singh, 2008. The midwives with whom I worked usually delayed washing the infants until several hours after birth because of the perceived advantages of leaving the vernix on. It was very common for the mothers to protest, however, and many demanded that the babies be washed as soon as possible.
26. Salk, 1960.
27. Miranda, 1970.
28. Trevathan, 1982.
29. Klaus and Kennell, 1976.
30. Dean Falk has argued that motherese may be the foundation of human language (Falk, 2004a; 2009).
31. Brazelton, 1963.
32. Trevathan, 1987; Klaus and Kennell, 1976.
33. Trevathan, 1987.
34. Hagen, 1999.
35. Wei et al., 2008.
36. Oates et al., 2004.
37. Hagen, 1999.
38. Hagen, 1999.
39. Hagen, 1999.
40. Scott, Klaus, and Klaus, 1999; Dennis, 2004.

41. Nesse, 1991.
42. Piperata, 2008.
43. Hrdy, 1999.
44. Hrdy, 1999.
45. Field, 1984; Edhborg et al., 2001; Murray, Fiori-Cowley, and Hooper, 1996; Beck, 1995; Cornish, McMahon, and Ungerer, 2008.
46. Hellin and Waller, 1992; Cooper, Murray, and Stein, 1993.
47. Corwin and Pajer, 2008; Finch, 2007.
48. Maes, et al., 2000.
49. Kendall-Tackett, 2007.
50. Kendall-Tackett, 2007.
51. Depression during late pregnancy increases the risk of premature birth (Dayan et al., 2006) and it appears that inflammation may be the proximate cause here, as well.
52. McEwan, 2003.
53. Kendall-Tackett, 2007. Instead of an apple a day, perhaps an ibuprofen a day is the best preventive new mothers can use!
54. McKenna and McDade, 2005.
55. Sleep deprivation is not necessarily related to negative mood, of course. Some mothers are so pleased to have a new baby that they cannot sleep for several days even though they may be in extremely good moods. Not surprisingly, medical attention has been focused primarily on negative mood so we tend to ignore the extremely positive moods that often follow birth.

Chapter 7

1. Neville, Morton, and Umemura, 2001.
2. Neville, 2001.
3. Goldman, 2001.
4. McKenna and McDade, 2005.
5. Wright, 2001.
6. Wright, 2001.
7. Rafael, 1973.
8. Dewey et al., 1992.
9. Waterlow and Thomson, 1979.
10. Berry and Gribble, 2008.
11. WHO Multicentre Growth Reference Study Group, 2006, page 82.
12. Karaolis-Danckert et al., 2007.
13. Hasselbalch et al., 1996.
14. Hasselbalch et al., 1999.
15. McDade, 2005.
16. American Academy of Pediatrics, 1997.
17. Wolf, 2006, page 400.
18. Merriam-Webster online dictionary.
19. Trevathan, 1987, page 32.

20. Picciano, 2001.
21. Sellen, 2006.
22. Anderson, Johnstone, and Remley, 1999.
23. Der, Batty, and Deary, 2006.
24. Jacobson and Jacobson, 2006.
25. Reynolds, 2001.
26. Pettitt et al., 1997.
27. Young et al., 2002.
28. Harder et al., 2005.
29. Simmons, 1997.
30. Harit et al., 2008.
31. Martin et al., 2005.
32. Martin, Gunnell, and Smith, 2005.
33. Martin et al., 2005, page 24.
34. McDade, 2005.
35. Goldman et al., 1998.
36. Goldman et al., 1998.
37. Heinig, 2001.
38. WHO World Health Statistics, 2008; McDade and Worthman, 1998.
39. Dettwyler, 1994.
40. Dell and To, 2001.
41. Oddy, 2004; Oddy et al., 2004.
42. McDade and Worthman, 1999.
43. Uvnas-Moberg, 1989.
44. Uvnas-Moberg, 1996.
45. Small, 1998; Barr, 1999; Schön, 2007.
46. Thompson, Olson, and Dessureau, 1996.
47. Barr, 1999.
48. Trevathan, 1987.
49. McKenna and McDade, 2005; Schön, 2007; Soltis, 2004.
50. Barr, 1999.
51. Schön, 2007; Papousek and Papousek, 1990; Jenni (2004) proposes that infant crying is tied to sleep-wake cycles and agrees that the incessant crying that is deemed unhealthy is often due to biocultural processes and parental behaviors that affect alignment of these cycles.
52. Soltis, 2004; Douglas, 2005.
53. Gill, White, and Anderson, 1984.
54. Douglas, 2005.
55. Gill et al., 1984, page 892.
56. Bowlby, 1969.
57. Blass, 2004, page 461.
58. Soltis, 2004.
59. Bard, 2004.
60. Blass, 2004, page 460.
61. Falk, 2004b.
62. Uvnas-Moberg, 1996.

63. Labbok, 2001.
64. Heinig and Dewey, 1997; weight loss after pregnancy may be highly desirable in health-rich nations like the United States, Canada, and those of Western Europe, but it is not necessarily a good thing in many populations and it was probably not welcomed in the distant past.
65. Uvnas-Moberg, 1996.
66. Sellen, 2006.
67. Uvnas-Moberg, 1989.
68. Stuebe, 2007.
69. Stuebe, 2007.
70. Stuebe et al., 2006.
71. Labbok, 2001.
72. Labbok, 2001.
73. Collaborative Group on Hormonal Factors in Breast Cancer, 2002.
74. Rosenblatt, Thomas, and The WHO Collaborative Study of Neoplasia and Steroid Contraceptives, 1995.
75. Sellen, 2006; McDade and Worthman, 1998; Tracer, 2002.
76. McDade and Worthman, 1998, page 296.
77. Li et al., 2008.
78. Grummer-Strawn et al., 2008.
79. Obermeyer and Castle, 1997.
80. Cole, Paul, and Whitehead, 2002.
81. A common reason for terminating breastfeeding is a woman's need to return to work, but working outside the home is not just a modern and Western issue. Of course women have taken on heavy workloads throughout history. Returning to working in the fields or foraging for food has a direct impact on energy balance and on rates and duration of breastfeeding. These, in turn, have an impact on reproduction. Where workloads are high, resources are constrained, and breastfeeding is prolonged, birth intervals are longer, but if any one of these factors is altered, reproductive patterns are affected (Panter Brick and Pollard, 1999).
82. Li et al., 2008.
83. Neville et al., 2001.
84. Dettwyler, 1995a.
85. Mintz, 2009.
86. kellymom, 2009.
87. Riordan, 1997.
88. Ryan and Zhou, 2006.
89. Lindsay, 2001.
90. Ball and Wright, 1999, page 870.
91. Dettwyler, 1995a; Kelleher, 2006.
92. Kelleher, 2006.
93. Wolf, 2006.
94. Kronberg and Væth, 2009.
95. Labbok, 2001.
96. Lawrence and Lawrence, 2001.

Chapter 8

1. Falk, 2009.
2. Wall-Scheffler, Geiger, and Steudel-Numbers, 2007.
3. Lee, 1980.
4. Tracer, 2009.
5. McKenna and McDade, 2005, page 135.
6. McKenna and McDade, 2005, page 136.
7. McKenna and McDade, 2005.
8. McKenna, 1986; McKenna, Mosko, and Richard, 1999; McKenna, 1996; McKenna, Ball, and Gettler, 2007.
9. McKenna and McDade, 2005, page 136.
10. Horne et al., 2002.
11. Ball and Klingaman, 2008.
12. McKenna and McDade, 2005.
13. McKenna, 1986; Lipsitt and Burns, 1986.
14. Laitman, 1984.
15. Laitman, 1986, page 66.
16. Laitman, 1984.
17. McKenna and McDade, 2005.
18. Sellen, 2006.
19. Dettwyler, 1995b.
20. Smith, 1991.
21. Kennedy, 2005.
22. Kennedy, 2005.
23. Sear and Mace, 2005.
24. Sellen, 2006; Lancaster and Lancaster, 1983; Bogin, 1998.
25. Bogin, 1998.
26. Sellen, 2006.
27. Sellen, 2007.
28. Lepore, 2009.
29. Lepore, 2009, page 39.
30. Wolf, 2007.
31. Gierson, 2002, page 78.
32. Kitzinger, 2009.
33. Trevathan and McKenna, 1994.
34. Christakis, 2008.
35. Falk, 2009.
36. Hrdy, 1999; Winnicott, 1964.

Chapter 9

1. Walker and Herndon, 2008.
2. Walker and Herndon, 2008.
3. Cohen, 2004.

4. Thompson et al., 2007.
5. Pavelka and Fedigan, 1991.
6. Nishida, 1997.
7. Pavelka and Fedigan, 1991.
8. Hawkes, 2003.
9. Lee, 1968.
10. Pavelka and Fedigan, 1991.
11. Hawkes, 2003.
12. Williams, 1957.
13. Peccei, 1995.
14. Mayer, 1982.
15. Hill and Hurtado, 1991.
16. Peccei, 2001.
17. If difficult birth is an important component in the evolution of menopause, it may explain why menopause is not common in most mammals, who also do not typically have such difficult births accompanied by high mortality (Perls and Fretts, 2001).
18. Perls and Fretts, 2001.
19. Temmermana et al., 2004.
20. Shanley and Kirkwood, 2001.
21. Voland, Chasiotis, and Schiefenhövel, 2005.
22. Mace and Sear, 2005.
23. Mace and Sear, 2005; Lee, 2003.
24. Sievert, 2006; Leidy, 1996.
25. Stanford et al., 1987.
26. Sievert, Waddle, and Canali, 2001.
27. Sievert, 2006.
28. Nagata et al., 1998, 1999.
29. Wilbur et al., 1990.
30. Whiteman et al., 2003; Gold et al., 2000.
31. Brooke and Long, 1987; Morgan, 1982.
32. Bachman, 1990.
33. McCoy, Cutler, and Davidson, 1985.
34. Whiteman et al., 2003.
35. Gold et al., 2000.
36. Birkhauser, 2002; Schindler, 2002.
37. Hunter, 1988, 1996; Nicol-Smith, 1996; Matthews et al., 1990.
38. Mitchell and Woods, 1996.
39. Kuh et al., 2002.
40. Schmidt, Haq, and Rubinow, 2004.
41. Gold et al., 2000.
42. See Robinson, 1996, for a review.
43. Sukwatana, et al., 1991.
44. Holte and Mikkelsen, 1991.
45. Prakash, Murthy, and Vinoda, 1982.
46. Okonofua, Lawal, and Bamgbose, 1990.
47. Moore and Kombe, 1991.

48. Lock, 1993.
49. Wright, 1983.
50. Beyenne and Martin, 2001.
51. George, 1988.
52. Williams, 1957.
53. Pavelka and Fedigan, 1991; Leidy, 1999.
54. Leidy, 1999; Hall, 2004.
55. Lock and Kaufert, 2001.
56. Pavelka and Fedigan, 1991.
57. Bell, 1987.
58. Kaufert and Lock, 1992.
59. Many people have the mistaken idea that most people in the ancestral past died before the end of their reproductive years so that to speak of menopause and senescence in the past is to discuss life phases that were very rare. It is true that infant and child mortality were quite high in the past (estimates are about 200 deaths per thousand) and that the *average* life expectancy may have been around 25–30 years (Cohen, 1989), but once they survived childhood, many people lived well into their 50s, 60s, and even 70s, based on demographic studies of nonagricultural populations.
60. Leidy, 1999.
61. Martin, 1987.
62. Derry, 2006.
63. Kaufert and Lock, 1992, page 204.
64. Parlee, 1990.
65. Mitchell and Woods, 1996; Avis et al., 2001.
66. Gracia and Freeman, 2004.
67. Avis et al., 2001.
68. Freeman et al., 2001.
69. Sievert, 2006.
70. Ellison, 2001; Pollard, 2008.
71. Derry, 2006.
72. Beyenne and Martin, 2001.
73. Galloway, 1997.
74. Galloway, 1997.
75. Galloway, 1997, page 144.
76. Campbell and Whitehead, 1977.
77. In our research, we used these terms to describe the behaviors we were interested in rather than *symptoms*, a term that reinforces the notion that menopause is a disease or disorder.
78. Burleson, Todd, and Trevathan, 2010.
79. Worthman and Melby, 2002; Worthman, 2008.
80. Worthman, 2008, page 300.
81. Worthman, 2008, page 310.
82. When I lived alone my cats often slept with me. I slept much better with them on the bed because I knew that they would quickly respond to an unfamiliar noise and I would be alerted by their movement. On the other hand, if I heard a noise and the cats weren't aroused, I knew it was not worrisome and went back to sleep. In this

way, cats are much better "watchdogs" than dogs who will often bark at their own shadows.

83. Worthman, 2008.
84. Sleep duration has recently been linked with obesity and type 2 diabetes (Pollard, 2008).
85. Worthman, 2008.
86. Counts, 1992; Hotvedt, 1991; Beyenne, 1989.
87. Shostak, 1981.
88. Bolin and Whelehan, 1999.
89. Postmenopausal women in health-rich nations have higher levels of estrogen than postmenopausal women in health-poor populations, which may be another contributor to high rates of breast cancer in older women (Pollard, 2008).
90. Austad, 1994.
91. Kerns and Brown, 1992.
92. Brown, 1992.
93. Lee, 1992.
94. Vatuk, 1992.
95. Lancaster and King, 1992.
96. Worthman, 1995.

Chapter 10

1. Blurton Jones, Hawkes, and O'Connell, 2002.
2. Hrdy, 1999.
3. Hall, 2004.
4. Paul, 2005.
5. Pavelka, Fedigan, and Zohar, 2002.
6. Hawkes, O'Connell, and Blurton Jones, 1997.
7. Hawkes et al., 1997.
8. Gurven and Hill, 1997.
9. Hawkes et al., 1997.
10. Gibson and Mace, 2005; Leonetti et al., 2007; Sear and Mace, 2008.
11. Shanley et al., 2007.
12. Lahdenperä et al., 2004.
13. Voland and Beise, 2002.
14. Madrigal and Meléndez-Obando, 2008.
15. Milne, 2006.
16. Many of these arguments pertain to long-lived grandfathers as well, but because my emphasis in this book is women's health, I focus more specifically on grandmothers.
17. Caspari and Lee, 2004.
18. Rosenberg, 2004.
19. Kirkwood, 2008.
20. Kirkwood, 2008.
21. Kirkwood, 2008.
22. Kaplan and Robson, 2002.
23. Kaplan and Robson, 2002; Kaplan et al., 2000.

24. Lee, 2003.
25. Kaplan and Robson, 2002.
26. Sorenson Jamison, Jamison, and Cornell, 2005.
27. Allen, Bruce, and Damasio, 2005.
28. Finch and Sapolsky, 1999.
29. Finch and Sapolsky, 1999.
30. Finch and Stanford, 2004.
31. Finch and Stanford, 2004.
32. Kaplan and Robson, 2002, page 10225.
33. Finch and Stanford 2004; Fullerton et al., 2000.
34. Finch and Morgan, 2007.
35. Finch and Stanford, 2004.
36. From an evolutionary perspective, it is important to emphasize that these two alleles differ from each other by only 2 amino acids in a sequence of 299. In other words, only very slight changes at the molecular level can have far-reaching physiological consequences, just as is seen in hemoglobin variants, which differ by only one amino acid.
37. Ewald, 2008.
38. Finch and Stanford, 2004.
39. Carstensen and Löckenhoff, 2003.
40. Carstensen and Löckenhoff, 2003.
41. Carstensen and Löckenhoff, 2003, page 155.
42. Carstensen and Löckenhoff, 2003.
43. Carstensen and Löckenhoff, 2003, page 167.
44. Williams et al., 2006.
45. Windsor, Anstey, and Rodgers, 2008.
46. Lee, 2003.
47. Hrdy, 1999; Uvnas-Moberg, 1998.
48. Again, much of this section could pertain equally to grandfathers.
49. Goodman, 2003.
50. Musil, 2000.
51. Sands, Goldberg-Glen, and Thornton, 2005.
52. Of course, as emphasized by Hrdy (1999), when examined on the individual family level, care by even a stressed-out grandmother whose health and well-being are compromised is better than most of the alternatives that would be available to the children; we always need to consider not just the optimum but the best of perhaps a number of bad or less-than-optimum alternatives.
53. Sorenson Jamison et al., 2005, pages 106–107.
54. Moalem, 2007.
55. Gurven et al., 2008.

Chapter 11

1. Moalem, 2007.
2. Vitzthum, 1997, page 257.
3. Ellison, 1999; Whitten, 1999.

4. Whitten, 1999.
5. Whitten, 1999.
6. Ellison, 1999.
7. Ewald, in press.
8. Note that the relationship is one of association, not of causation. The links are still being explored and causes of chronic diseases are far more complex than any single "cause" we might search for.
9. HIV/AIDS is the classic and most unfortunate example. Many of the people who acquire HIV have full knowledge of what causes the disease, but they either do not have the power to make changes (for example, in the case of married women who have little or no control over their husband's sexual behavior) or continue engaging in risky behavior despite the knowledge. In a study we did of commercial and freelance sex workers in the Philippines, almost all of the women could correctly answer questions about methods of preventing AIDS and other STDs but most did not practice "safe sex" behaviors (Amadora-Nolasco et al., 2001). For many, earning money in the sex trade was the only way they could support their children—more trade-offs. The risk of contracting AIDS was less real to them than the risk to their children's health if they adopted safe sex practices that interfered with their ability to earn money.
10. Schooling and Kuh, 2002.
11. Schooling and Kuh, 2002, page 289.
12. Shubin, 2008.
13. Kuh, dos Santos Silva, and Barrett-Connor, 2002.
14. Total population for the Philippines in 2008 was estimated to be about 96 million.
15. Worthman, 1999.
16. For exceptions, see Eaton, Shostak, and Konner, 1988; Nathanielsz, 2001.
17. Dos Santos Silva and de Stavola, 2002.

References

Abitbol, M. Maurice. 1987a. Evolution of the sacrum in hominoids. *American Journal of Physical Anthropology* 74: 65–81.

Abitbol, M. Maurice. 1987b. Obstetrics and posture in pelvic anatomy. *Journal of Human Evolution* 16: 243–255.

Abitbol, M. Maurice. 1988. Evolution of the ischial spine and of the pelvic floor in the Hominoidea. *American Journal of Physical Anthropology* 75: 53–67.

Abitbol, M. Maurice. 1993. Growth of the fetus in the abdominal cavity. *American Journal of Physical Anthropology* 91: 367–378.

Ackerman, Jennifer. 2006. The downside of upright. *National Geographic*, July, pp. 126–145.

Adair, L. S., Kuzawa, C. W. 2001. Early growth retardation and syndrome X: Conceptual and methodological issues surrounding the programming hypothesis. In *Nutrition and Growth*, ed. R. Martorell, F. Haschke, pp. 333–350. Geneva: Nestle Nutrition Workshop Series.

Aiello, Leslie C., Wheeler, Peter. 1995. The expensive-tissue hypothesis: The brain and the digestive system in human and primate evolution. *Current Anthropology* 36: 199–221.

Allen, John S., Bruss, Joel, Damasio, Hanna. 2005. The aging brain: The cognitive reserve hypothesis and hominid evolution. *American Journal of Human Biology* 17: 673–689.

Allman, John, Hasenstaub, Andrea. 1999. Brains, maturation times, and parenting. *Neurobiology of Aging* 20: 447–454.

Amadora-Nolasco, Fiscalina, Alburo, René E., Aguilar, Elmira Judy T., Trevathan, Wenda R. 2001. Knowledge, perception of risk for HIV and condom use: A comparison of establishment-based and freelance female sex workers in Cebu City, Philippines. *AIDS and Behavior* 4: 319–330.

American Academy of Pediatrics. 1997. Breastfeeding and the use of human milk. *Pediatrics* 100: 1035–1039.

Anderson, James W., Johnstone, Bryan M., Remley, Daniel T. 1999. Breast-feeding and cognitive development: A meta-analysis. *American Journal of Clinical Nutrition* 70: 525–535.

Anderson, M. 1971. Transplantation—Nature's success. *Lancet* 2: 1077–1082.

Angold, A., Costello, E. J., Worthman, C. M. 1998. Puberty and depression: The roles of age, pubertal status and pubertal timing. *Psychological Medicine* 28: 51–61.

Angstwurm, M. W. A., Gartner, R., Ziegler-Heitbrock, H. W. L. 1997. Cyclic plasma Il-6 levels during normal menstrual cycle. *Cytokine* 9: 370–374.

Apter, D., Vihko, R. 1983. Early menarche, a risk factor for breast cancer, indicates early onset of ovulatory cycles. *Journal of Clinical Endocrinology & Metabolism* 57: 82–86.

Armelagos, G. J., Brown, P. J., Turner, B. 2005. Evolutionary, historical and political economic perspectives on health and disease. *Social Science & Medicine* 61: 755–765.

Austad, S. N. 1994. Menopause: An evolutionary perspective. *Experimental Gerontology* 29: 255–263.

Avis, N. E., Stellato, R., Crawford, S. L., Bromberger, J. T., Ganz, P., Cain, V., Kagawa-Singer, M. 2001. Is there a menopausal syndrome? Menopausal status and symptoms across racial/ethnic groups. *Social Science and Medicine* 52: 345–356.

Bachmann, G. A. 1990. Sexual issues at menopause. *Annals of the New York Academy of Sciences* 592: 87–93.

Baker, Nena. 2008. *The Body Toxic*. New York: North Point Press.

Ball, Helen, Klingaman, Kristin. 2008. Breastfeeding and mother-infant sleep proximity. In *Evolutionary Medicine and Health: New Perspectives*, ed. Wenda R. Trevathan, E. O. Smith, James J. McKenna, pp. 226–241. New York: Oxford University Press.

Ball, Thomas M., Wright, Anne L. 1999. Health care costs of formula-feeding in the first year of life. *Pediatrics* 103: 870–876.

Barash, David P. 2005. Does God have back problems too? *Los Angeles Times*, June 27, pp. B-9.

Barber, Nigel. 2001. Marital opportunity, parental investment, and teen birth rates of blacks and whites in American states. *Cross-Cultural Research* 35: 263–279.

Barber, Nigel. 2002. Parental investment prospects and teen birth rates of blacks and whites in American metropolitan areas. *Cross-Cultural Research* 36: 183–199.

Bard, Kim A. 2004. What is the evolutionary basis for colic? *Behavioral and Brain Sciences* 27: 459.

Barker D. J. 1995. Fetal origins of coronary heart disease. *British Medical Journal* 311: 171–174.

Barker, D. J. P. 1997. Fetal nutrition and cardiovascular disease in later life. *British Medical Bulletin* 53: 96–108.

Barker, D. J. P. 1998. *Mothers, Babies, and Health in Later Life*. Edinburgh, UK: Churchill Livingstone.

Barker, D. J. P. 1999. The fetal origins of type 2 diabetes mellitus. *Annals of Internal Medicine* 130: 322–324.

Barker, D. J. P. 2004. The developmental origins of adult disease. *Journal of the American College of Nutrition* 23: 588S–595S.

Barker, D. J. P. 2005. The developmental origins of insulin resistance. *Hormone Research* 64: 2–7.

Barker, D. J., Bull, A. R., Osmond, C., Simmonds, S. J. 1990. Fetal and placental size and risk of hypertension in adult life. *British Medical Journal* 301: 259–263.

Barker, D. J. P., Osmond, C., Golding, J., Kuh, D., Wadsworth, M. E. 1989. Growth in utero, blood pressure in childhood and adult life, and mortality from cardiovascular disease. *British Medical Journal* 298: 564–567.

Barker, David J. P., Osmond, Clive, Thornburg, Kent L., Kajantie, Eero, Forsen, Tom J. 2008a. A possible link between the pubertal growth of girls and breast cancer in their daughters. *American Journal of Human Biology* 20: 127–131.

Barker, David J. P., Osmond, Clive, Thornburg, Kent L., Kajantie, Eero, Eriksson, Johan G. 2008b. A possible link between the pubertal growth of girls and ovarian cancer in their daughters. *American Journal of Human Biology* 20: 659–662.

Barnett, D. K, Abbott, D. H. 2003. Reproductive adaptations to a large-brained fetus open a vulnerability to anovulation similar to polycystic ovary syndrome. *American Journal of Human Biology* 15: 296–319.

Barr, Ronald G. 1999. Infant crying behavior and colic: An interpretation in evolutionary perspective. In *Evolutionary Medicine*, ed. W. R. Trevathan, E. O. Smith, J. J. McKenna, pp. 27–52. New York: Oxford University Press.

Barrett, Ronald, Kuzawa, Christopher W., McDade, Thomas, Armelagos, George J. 1998. Emerging infectious disease and the third epidemiological transition. In *Annual Review Anthropology*, pp. 247–271. Palo Alto: Annual Reviews Inc.

Baxley, Elizabeth G., Gobbo, Robert W. 2004. Shoulder dystocia. *American Family Physician* 69: 1707–1714.

Bdolah, Y., Sukhatme, V. P., Karumanchi, S. A., Bdolah, Yuval, Sukhatme, Vikas P., Karumanchi, S. Ananth. 2004. Angiogenic imbalance in the pathophysiology of preeclampsia: Newer insights. *Seminars in Nephrology* 24: 548–556.

Beck, C. T. 1995. The effects of postpartum depression on maternal-infant interaction: A meta-analysis. *Nursing Research* 44: 298–304.

Beer, E., Folghera, M. G. 2002. Early history of McRoberts' maneuver. *Minerva Ginecologica* 54: 197–199.

Beer, Eugenio. 2003. A guest editorial: Shoulder dystocia and posture for birth: A history lesson. *Obstetrical & Gynecological Survey* 58: 697–699.

Bell, Susan E. 1987. Changing ideas: The medicalization of menopause. *Social Science & Medicine* 24: 535–542.

Belsky, J., Steinberg, L., Draper, P. 1991. Childhood experience, interpersonal development, and reproductive strategy: An evolutionary theory of socialization. *Child Development* 62: 647–670.

Bener, A., Denic, S., Galadari, S. 2001. Longer breast-feeding and protection against childhood leukaemia and lymphomas. *European Journal of Cancer* 37: 234–238.

Bentley, G. R. 1994. Ranging hormones: Do hormonal contraceptives ignore human biological variation and evolution? *Annals of the New York Academy of Sciences* 709: 201–203.

Bentley, Gillian R. 1999. Aping our ancestors: comparative aspects of reproductive ecology. *Evolutionary Anthropology* 7: 175–185.

Berkey, Catherine S., Gardner, Jane D., Frazier, A. Lindsay, Colditz, Graham A. 2000. Relation of childhood diet and body size to menarche and adolescent growth in girls. *American Journal of Epidemiology* 152: 446–452.

Berry, Nina J., Gribble, Karleen D. 2008. Breast is no longer best: Promoting normal infant feeding. *Maternal and Child Nutrition* 4: 74–79.

Beyene, Y. 1989. *From Menarche to Menopause*. Albany: State University of New York Press.

Beyene, Yewoubdar, Martin, Mary C. 2001. Menopausal experiences and bone density of Mayan women in Yucatan, Mexico. *American Journal of Human Biology* 13: 505–511.

Birkhauser, M. 2002. Depression, menopause and estrogens: Is there a correlation? *Maturitas* 41: S3–S8.

Bjorkland, David F. 1997. The role of immaturity in human development. *Psychological Bulletin* 122: 153–169.

Blass, Elliott M. 2004. Changing brain activation needs determine early crying: A hypothesis. *Behavioral and Brain Sciences* 27: 460–461.

Blurton Jones, N. G. 1978. Natural selection and birthweight. *Annals of Human Biology* 5: 487–489.

Blurton Jones, N.G., Hawkes, Kristen, O'Connell, James F. 2002. Antiquity of postreproductive life: Are there modern impacts on hunter-gatherer postreproductive life spans? *American Journal of Human Biology* 14: 184–205.

Bogin, B. 1998. Evolutionary and biological aspects of childhood. In *Biosocial Perspectives on Children*, ed. C. Panter-Brick, pp. 10–44. Cambridge, UK: Cambridge University Press.

Bogin, B., Loucky, J. 1997. Plasticity, political economy, and physical growth status of Guatemala Maya children living in the United States. *American Journal of Physical Anthropology* 102: 17–32.

Bogin, B., Smith, B. H. 1996. Evolution of the human life cycle. *American Journal of Human Biology* 8: 703–716.

Bogin, B., Smith, P., Orden, A. B., Varela-Silva, M. I., Loucky, J. 2002. Rapid change in height and body proportions of Mayan American children. *American Journal of Human Biology* 14: 753–761.

Bogin, B., Varela-Silva, M. I., Rios, L. 2007. Life history trade-offs in human growth: Adaptation or pathology? *American Journal of Human Biology* 19: 631–642.

Bolin, Anne, Whelehan, Patricia. 1999. *Perspectives on Human Sexuality*. Albany: State University of New York Press.

Borgerhoff Mulder, Monique. 1989. Menarche, menopause, and reproduction in the Kipsigis of Kenya. *Journal of Biosocial Sciences* 21: 179–192.

Bowlby, John. 1969. *Attachment and Loss: Volume 1. Attachment*. New York: Basic Books.

Brandt, I. 1998. Neurological development. In *Cambridge Encyclopedia of Human Growth and Development*, ed. S. J. Ulijaszek, F. E. Johnston, M. A. Preece, pp. 164–165. Cambridge, UK: Cambridge University Press.

Brazelton, T. B. 1963. The early mother-infant adjustment. *Pediatrics* 32: 931–938.

Brewis, Alexandra, Meyer, Mary. 2005. Demographic evidence that human ovulation is undetectable (at least in pair bonds). *Current Anthropology* 46: 465–471.

Bribiescas, Richard. G., Ellison, Peter T. 2008. How hormones mediate trade-offs in human health and disease. In *Evolution in Health and Disease*, ed. Stephen C. Stearns, Jacob C. Koella, pp. 77–94. Oxford: Oxford University Press.

Bronson, F. H. 2000. Puberty and energy reserves: A walk on the wild side. In *Reproduction in Context*, ed. K. Wallen, J. E. Schneider. Cambridge, MA: MIT Press.

Brooke, S. T., Long, B. C. 1987. Efficiency of coping with a real-life stressor: A multimodal comparison of aerobic fitness. *Psychophysiology* 24: 173–180.

Brown, Judith K. 1992. Lives of middle-aged women. In *In Her Prime*, ed. Virginia Kerns, Judith K. Brown, pp. 17–30. Urbana: University of Illinois Press.

Brown, P. J., Konner, M. J. 1987. An anthropological perspective on obesity. *Annals of the New York Academy of Science* 499: 29–46.

Brown, Susan G., Powell, Grace E., Germone, Tamra J. 1997. The relation between phase of menstrual cycle and health related symptoms: An evolutionary perspective. *Advances in Ethology* 32: 67.

Buckley, Thomas, Gottlieb, Alma, eds. 1988. *Blood Magic: The Anthropology of Menstruation*. Berkeley: University of California Press.

Burger, J., Gochfeld, M. 1985. A hypothesis on the role of pheromones on age of menarche. *Medical Hypotheses* 17: 39–46.

Burleson, M. H., Trevathan, Wenda R., Gregory, W. L. 2002. Sexual behavior in lesbian and heterosexual women: Effects of menstrual cycle phase and partner availability. *Psychoneuroendocrinology* 27: 489–503.

Burleson, Mary H., Todd, Michael, Trevathan, Wenda R. 2010. Daily vasomotor symptoms, sleep problems, and mood: Using daily data to evaluate the domino hypothesis in mid-aged women. *Menopause* 17: 87–95.

Butler, H. 1977. The effect of thalidomide on a prosimian: The greater galago. *Journal of Medical Primatology* 6: 319–324.

Bystrova, Ksenia. 2009. Novel mechanism of human fetal growth regulation: A potential role of lanugo, vernix caseosa and a second tactile system of unmyelinated low-threshold C-afferents. *Medical Hypotheses*72: 143–146.

Campbell, B. C., Lukas, W. D., Campbell, K. L. 2001. Reproductive ecology of male immune function and gonadal function. In *Reproductive Ecology and Human Evolution*, ed. Peter T. Ellison, pp. 159–178. New York: Aldine de Gruyter.

Campbell, S., Whitehead, M. 1977. Oestrogen therapy and the menopausal syndrome. *Clinical Obstetrics and Gynaecology* 4: 31–47.

Cardwell, C. R., Stene, L. C., Joner, G., Cinek, O., Svensson, J., Goldacre, M. J., Parslow, R. C., Pozzilli, P., Brigis, G., Stoyanov, D., Urbonaitė, B., Šipetić, S., Schober, E., Ionescu-Tirgoviste, C., Devoti, G., de Beaufort, C. E., Buschard, K., Patterson, C. C. 2008. Caesarean section is associated with an increased risk of childhood-onset type 1 diabetes mellitus: A meta-analysis of observational studies. *Diabetologia* 51: 726–735.

Carstensen, Laura L., Löckenhoff, Corinna E. 2003. Aging, emotion, and evolution: The bigger picture. *Annals of the New York Academy of Science* 1000: 152–179.

Caspari, R., Lee, S. H. 2004. Older age becomes common late in human evolution. *Proceedings of the National Academy of Sciences of the United States of America* 101: 10895–10900.

Chisholm, James S. 1993. Death, hope, and sex: Life-history theory and the development of reproductive strategies. *Current Anthropology* 34: 1–24.

Chisholm, James S., Coall, David A. 2008. Not by bread alone: The role of psychosocial stress in age at first reproduction and health inequalities. In *Evolutionary Medicine and Health: New Perspectives*, ed. Wenda R. Trevathan, E. O. Smith, James J. McKenna, pp. 134–148. New York: Oxford University Press.

Chong, Yap-Seng, Liang, Yu, Gazzard, Guss, Stone, Richard A., Saw, Seang-Mei. 2005. Association between breastfeeding and likelihood of myopia in children. *Journal of the American Medical Association* 293: 3001–3002.

Christakis, Dimitri A. 2008. The effects of infant media usage: What do we know and what should we learn? *Acta Pædiatrica* 98: 8–16.

Chung, C. S., Morton, N. E. 1961. Selection at the ABO locus. *American Journal of Human Genetics* 13: 9–27.

Clancy, K. B. H., Nenko, I., Jasienska, G. 2006. Menstruation does not cause anemia: Endometrial thickness correlates positively with erythrocyte count and hemoglobin concentration in premenopausal women. *American Journal of Human Biology* 18: 710–713.

Clapp, James F. III, Rokey, Roxanne, Treadway, Judith L., Carpenter, Marshall W., Artal, Raul M., Warrnes, Carole. 1992. Exercise in pregnancy. *Medicine and Science in Sports and Exercise* 24: S294–S300.

Coall, D. A., Chisholm, J. S. 2003. Evolutionary perspectives on pregnancy: Maternal age at menarche and infant birth weight. *Social Science & Medicine* 57: 1771–1781.

Cohen, Alan A. 2004. Female post-reproductive lifespan: A general mammalian trait. *Biological Reviews* 79: 733–750.

Cohen, Joel E. 2003. Human population: The next half century. *Science* 302: 1172–1175.

Cohen, Mark N. 1989. *Health and the Rise of Civilization.* New Haven: Yale University Press.

Cole, T. J., Paul, A. A., Whitehead, R. G. 2002. Weight reference charts for British long-term breastfed infants. *Acta Pædiatrica* 91: 1296–1300.

Colhoun, Helen M., Chaturvedi, Nish. 2002. A life course approach to diabetes. In *A Life Course Approach to Women's Health,* ed. Diana Kuh, Rebecca Hardy, pp. 121–140. Oxford: Oxford University Press.

Collaborative Group on Hormonal Factors in Breast Cancer. 2002. Breast cancer and breast feeding: Collaborative reanalysis of individual data from 47 epidemiological studies in 30 countries, including 50,302 women with breast cancer and 96,973 women without the disease. *Lancet* 360: 187–195.

Cooper, P. J., Murray, L., Stein, A. 1993. Psychosocial factors associated with early termination of breastfeeding. *Journal of Psychosomatic Research* 37: 171–176.

Cornish, Alison M., Mcmahon, Catherine, Ungerer, Judy A. 2008. Postnatal depression and the quality of mother-infant interactions during the second year of life. *Australian Journal of Psychology* 60: 142–151.

Corwin, Elizabeth J., Pajer, Kathleen. 2008. The psychoneuroendocrinology of postpartum depression. *Journal of Women's Health* 17: 1529–1534.

Costello, Elizabeth Jane, Worthman, Carol M., Erkanli, Alaattin, Angold, Adrian. 2007. Prediction from low birth weight to female adolescent depression. *Archives of General Psychiatry* 64: 338–344.

Counts, Dorothy Ayers. 1992. Tamparonga:"The Big Women" of Kaliai (Papua New Guinea). In *In Her Prime,* ed. Virginia Kerns, Judith K. Brown. Urbana: University of Illinois Press.

Coutinho, Elsimar M., Segal, Sheldon J. 1999. *Is Menstruation Obsolete?* New York: Oxford University Press.

Crews, Douglas E. 2003. *Human Senescence: Evolutionary and Biocultural Perspectives.* Cambridge, UK: Cambridge University Press.

Daly, Martin, Wilson, Margo. 1983. *Sex, Evolution, and Behavior.* Belmont, CA: Wadsworth.

Davis, Elysia Poggi, Hobel, Calvin J., Sandman, Curt A., Glynn, Laura M., Wadhwa, Pathik D. 2005. Prenatal stress and stress physiology influences human fetal and infant development. In *Birth, Distress, and Disease,* ed. M. L. Power, J. Schulkin, pp. 183–201. Cambridge, UK: Cambridge University Press.

Davis, Margarett K. 2001. Breastfeeding and chronic disease in childhood and adolescence. *Pediatric Clinics of North America* 48: 125–141.

Dayan, Jacques, Creveuil, Christian, Marks, Maureen N., Conroy, Sue, Herlicoviez, Michel, Dreyfus, Michel, Tordjman, Sylvie. 2006. Prenatal depression, prenatal anxiety, and spontaneous preterm birth: A prospective cohort study among women with early and regular care. *Psychosomatic Medicine* 68: 938–946.

de Benoist, Bruno, Andersson, Maria, Egli, Ines, Takkouche, Bahi, Allen, Henrietta. 2004. *Iodine Status Worldwide.* Geneva: World Health Organization.

Dell, Sharon, To, Theresa. 2001. Breastfeeding and asthma in young children. *Archives of Pediatric & Adolescent Medicine* 155: 1261–1265.

Dennis, Cindy-Lee E. 2004. Preventing postpartum depression part II: A critical review of nonbiological interventions. *Canadian Journal of Psychiatry* 49: 526–538.

Der, Geoff, Batty, G. David, Deary, Ian J. 2006. Effect of breast feeding on intelligence in children: Prospective study, sibling pairs analysis, and meta-analysis. *British Medical Journal* 333: 945–948.

Derry, Paula S. 2006. A lifespan biological model of menopause. *Sex Roles* 54: 393–399.

DeSilva, Jeremy M., Lesnik, Julie J. 2008. Brain size at birth throughout human evolution: A new method for estimating neonatal brain size in hominins. *Journal of Human Evolution* 55: 1064–1074.

Dettwyler, K. A. 1994. *Dancing Skeletons*. Prospect Heights, IL: Waveland Press.

Dettwyler, K. A. 1995a. Beauty and the breast: The cultural context of breastfeeding in the United States. In *Breastfeeding: Biocultural Perspectives*, ed. P. Stuart-Macadam, K. A. Dettwyler, pp. 167–216. New York: Aldine de Gruyter.

Dettwyler, K. A. 1995b. A time to wean: The hominid blueprint for the natural age of weaning in modern human populations. In *Breastfeeding: Biocultural Perspectives*, ed. P. Stuart-Macadam, K. A. Dettwyler, pp. 39–73. New York: Aldine de Gruyter.

Dewey, Kathryn G., Heinig, Jane, Nommsen, Laurie A., Peerson, Janet M, Lönnerdal, Bo. 1992. Growth of breast-fed and formula-fed infants from 0 to 18 months: The DARLING study. *Pediatrics* 89: 1035–1041.

Diamond, J. 1992. A question of size. *Discover* 13: 70–77.

Dietz, H. P. 2008. The aetiology of prolapse. *International Urogynecology Journal* 19: 1323–1329.

Doherty, Elizabeth G., Eichenwald, Eric C. 2004. Cesearean delivery: Emphasis on the neonate. *Clinical Obstetrics & Gynecology* 47: 332–341.

dos Santos Silva, I., de Stavola, B. 2002. Breast cancer aetiology: Where do we go from here? In *A Life Course Approach to Women's Health*, ed. Diana Kuh, Rebecca Hardy, pp. 44–63. Oxford: Oxford University Press.

Douglas, Pamela S. 2005. Excessive crying and gastro-oesophageal reflux disease in infants: Misalignment of biology and culture. *Medical Hypotheses* 64: 887–898.

Doyle, Caroline, Swain Ewald, Holly A., Ewald, Paul W. 2008. An evolutionary perspective on premenstrual syndrome: Implications for investigating infectious causes of chronic disease. In *Evolutionary Medicine and Health: New Perspectives*, ed. W. R. Trevathan, E. O. Smith, J. J. McKenna, pp. 196–215. New York: Oxford University Press.

Draper, Pat, Harpending, Henry. 1982. Father absence and reproductive strategies: An evolutionary perspective. *Journal of Anthropological Research* 38: 255–273.

Dufour, D. L., Sauther, M. L. 2002. Comparative and evolutionary dimensions of the energetics of human pregnancy and lactation. *American Journal of Human Biology* 14: 584–602.

Eaton, S. B., Eaton, S. B. III. 1999. Breast cancer in evolutionary perspective. In *Evolutionary Medicine*, ed. W. R. Trevathan, E. O. Smith, J. J. McKenna, pp. 429–442. New York: Oxford University Press.

Eaton, S. B., Pike, M. C., Short, R. V., Lee, N. C., Trussell, J., Hatcher, R. A., Wood, J. W., Worthman, C. M., Blurton Jones, N. G., Konner, M. J., Hill, K. R., Bailey, R., Hurtado, A. M. 1994. Women's reproductive cancers in evolutionary context. *Quarterly Review of Biology* 69: 353–367.

Eaton, S. B., Shostak, M., Konner, M. 1988. *The Paleolithic Prescription: A Program of Diet, Exercise and a Design for Living*. New York: Harper and Row.

Edhborg, M., Lundh, W., Seimyr, L., Widstrom, A-M. 2001. The long-term impact of postnatal depressed mood on mother-child interaction: A preliminary study. *Journal of Reproductive and Infant Psychology* 19: 61–71.

Ellis, Bruce J. 2004. Timing of pubertal maturation in girls: An integrated life history approach. *Psychological Bulletin* 130: 920–958.

Ellison, P. T. 1988. Human salivary steroids: Methodological considerations and applications in physical anthropology. *Yearbook of Physical Anthropology* 31: 115–142.

Ellison, P. T. 1990. Human ovarian function and reproductive ecology: New hypotheses. *American Anthropologist* 92: 933–952.

Ellison, P. T. 1994. Salivary steroids and natural variation in human ovarian function. *Annals of the New York Academy of Science* 709: 287–298.

Ellison, P. T. 1999. Reproductive ecology and reproductive cancers. In *Hormones, Health and Behaviour. A Socio-Ecological and Lifespan Perspective*, ed. C. Panter-Brick, C. Worthman, pp. 184–209. Cambridge, UK: Cambridge University Press.

Ellison, P. T. 2001. *On Fertile Ground: A Natural History of Human Reproduction*. Cambridge, MA: Harvard University Press.

Ellison, P. T. 2005. Evolutionary perspectives on the fetal origins hypothesis. *American Journal of Human Biology* 17: 113–118.

Ensom, Mary H. H. 2000. Gender-based differences and menstrual cycle-related changes in specific diseases: Implications for pharmacotherapy. *Pharmacotherapy* 20: 523–539.

Ewald, Paul W. 1994. *Evolution of Infectious Disease*. New York: Oxford University Press

Ewald, Paul W. 1999. Evolutionary control of HIV and other sexually transmitted viruses. In *Evolutionary Medicine*, ed. Wenda R. Trevathan, E. O. Smith, James J. McKenna, pp. 271–312. New York: Oxford University Press.

Ewald, Paul W. 2008. An evolutionary perspective on the causes of chronic diseases. In *Evolutionary Medicine and Health: New Perspectives*, ed. Wenda R. Trevathan, E. O. Smith, James J. McKenna, pp. 350–367. New York: Oxford University Press.

Ewald, Paul W. in press. Evolutionary medicine and the causes of chronic disease. In *Human Evolutionary Biology*, ed. Michael P. Muehlenbein. Cambridge, UK: Cambridge University Press.

Falk, Dean. 2004a. Prelinguistic evolution in early hominins: Whence motherese? *Behavioral and Brain Sciences* 27: 491–541.

Falk, Dean. 2004b. Prelinguistic evolution in hominin mothers and babies: For cryin' out loud! *Behavioral and Brain Sciences* 27: 461–462.

Falk, Dean. 2009. *Finding Our Tongues: Mothers, Infants, and the Origins of Language*. New York: Basic Books.

Fessler, D. M. T. 2002. Reproductive immunosuppression and diet: An evolutionary perspective on pregnancy sickness and meat consumption. *Current Anthropology* 43: 19–61.

Fidel, P. L., Cutright, J., Steele, C. 2000. Effects of reproductive hormones on experimental vaginal candidiasis. *Infection and Immunity* 68: 651–657.

Field, T. 1984. Early interactions between infants and their postpartum depressed mothers. *Infant Behavior and Development* 7: 517–522.

Finch, Caleb E. 2007. *The Biology of Human Longevity: Inflammation, nutrition, and aging in the evolution of life spans*. London: Academic Press.

Finch, Caleb E., Morgan, Todd E. 2007. Systemic inflammation, infection, apoE alleles, and Alzheimer disease: A position paper. *Current Alzheimer Research* 4: 185–189.

Finch, Caleb E., Rose, Michael R. 1995. Hormones and the physiological architecture of life history evolution. *Quarterly Review of Biology* 70: 1–52.

Finch, Caleb E., Sapolsky, R. M. 1999. The evolution of Alzheimer disease, the reproductive schedule, and apoE isoforms. *Neurobiology of Aging* 20: 407–428.

Finch, Caleb E., Stanford, Craig B. 2004. Meat-adaptive genes and the evolution of slower aging in humans. *Quarterly Review of Biology* 79: 3–50.

Finn, C. A. 1998. Menstruation: A nonadaptive consequence of uterine evolution. *Quarterly Review of Biology* 73: 163–73

Fisher, H. E. 1983. *The Sex Contract.* New York: Quill.

Flaxman, S. M., Sherman, P. W. 2000. Morning sickness: A mechanism for protecting mother and embryo. *Quarterly Review of Biology* 75: 113–148.

Flaxman, S. M., Sherman, P. W. 2008. Morning sickness: Adaptive cause or nonadaptive consequence of embryo viability? *American Naturalist* 172: 54–62.

Forbes, Scott. 2002. Pregnancy sickness and embryo quality. *Trends in Ecology & Evolution* 17: 115–120.

Freeman, E. W., Grisso, J. A., Berlin, J., Sammel, M. D., Garcia-Espana, B., Hollander, L. 2001. Symptom reports from a cohort of African American and white women in the late reproductive years. *Menopause* 8: 33–42.

Freudigman, Kimberly A., Thoman, E. B. 1998. Infants' earliest sleep/weak organization differs as a function of delivery mode. *Developmental Psychobiology* 32: 293–303.

Friede, Andrew, Baldwin, Wendy, Rhodes, Philip H., Buehler, James W., Strauss, Lilo T., Smith, Jack C., Hogue, Carol J. R. 1987. Young maternal age and infant mortality: The role of low birth weight. *Public Health Reports* 102: 192–199.

Frisch, R. E., McArthur, J. W. 1974. Menstrual cycles: Fatness as a determinant of minimum weight for height necessary for their maintenance or onset. *Science* 185: 949–951.

Frisch, R. E., Revelle, R. 1970. Height and weight at menarche and a hypothesis of critical body weights and adolescent events. *Science* 169: 397–399.

Frisch, R., Wyshak, G., Albright, N., Albright, T., Schiff, I., Jones, K., Witschi, J., Shiang, E., Koff, E., Marguglio, M. 1985. Lower prevalence of breast cancer and cancers of the reproductive system among former college athletes compared to non-athletes. *British Journal of Cancer* 52: 885–91.

Fullerton, S. M., Clark, A. G., Weiss, K. M., Nickerson, D. A., Taylor, S. L., Stengard, J. H., Salomaa, V., Vartiainen, E., Perola, M., Boerwinkle, E., Sing, C. F. 2000. Apolipoprotein E variation at the sequence haplotype level: Implications for the origin and maintenance of a major human polymorphism. *American Journal of Human Genetics* 67: 881–900.

Furneaux, E. C., Langley-Evans, A. J., Langley-Evans, S. C. 2001. Nausea and vomiting of pregnancy: Endocrine basis and contribution to pregnancy outcome. *Obstetrical and Gynecological Survey* 56: 775–782.

Galloway, Alison. 1997. The cost of reproduction and the evolution of postmenopausal osteoporosis. In *The Evolving Female: A Life-History Perspective,* ed. Mary Ellen Morbeck, Alison Galloway, Adrienne L. Zihlman. Princeton, NJ: Princeton University Press.

Gardner, J. 1983. Adolescent menstrual characteristics as predictors of gynaecological health. *Annals of Human Biology* 10: 31–40.

Garland, M., Doherty, D., Golden-Mason, L., Fitzpatrick, P., Walsh, N., O'Farrelly, C. 2003. Stress-related hormonal suppression of natural killer activity does not show menstrual cycle variation: Implication for timing of surgery for breast cancer. *Anticancer Research* 23: 2531–2535.

George, T. 1988. Menopause: Some interpretations of the results of a study among a non-western group. *Maturitas* 10: 109–116.

Geronimus, Arline T. 1987. On teenage childbearing and neonatal mortality in the United States. *Population and Development Review* 13: 245–279.

Geronimus, Arline T. 2003. Damned if you do: Culture, identity, privilege, and teenage childbearing in the United States. *Social Science & Medicine* 57: 881–893.

Gibson, Mhairi A., Mace, Ruth. 2005. Helpful grandmothers in rural Ethiopia: A study of the effect of kin on child survival and growth. *Evolution and Human Behavior* 26: 469–482.

Gierson, Bruce. 2002. The year in ideas: The crying-baby translator. *New York Times*, December 15, p. 78.

Gill, N. E., White, M. A., Anderson, G. C. 1984. Transitional newborn infant in a hospital nursery—from first oral cue to first sustained cry. *Nursing Research* 33: 213–217.

Gleicher, Norbert, Barad, David. 2006. An evolutionary concept of polycystic ovarian disease: Does evolution favour reproductive success over survival? *Reproductive BioMedicine Online* 12: 587–589.

Gluckman, Peter D., Hanson, Mark. 2006. *Mismatch: Why Our World No Longer Fits Our Bodies*. Oxford, UK: Oxford University Press.

Gluckman, Peter D., Hanson, Mark A., Beedle, Alan S. 2007. Early life events and their consequences for later disease: A life history and evolutionary perspective. *American Journal of Human Biology* 19: 1–19.

Glynn, Laura M., Wadhwa, Pathik D., Dunkel-Schetter, Christine, Chicz-DeMet, Aleksandra, Sandman, Curt A. 2001. When stress happens matters: Effects of earthquake timing on stress responsivity in pregnancy. *American Journal of Obstetrics and Gynecology* 184: 637–642.

Goer, Henci. 2001. The case against elective cesarean section. *Journal of Perinatal and Neonatal Nursing* 15: 23–38.

Gold, E. B., Sternfeld, B., Kelsey, J. L., Brown, C., Mouton, C., Reame, N., Salamone, L., Stellato, R. 2000. Relation of demographic and lifestyle factors to symptoms in a multi-racial/ethnic population of women 40–55 years of age. *American Journal of Epidemiology* 152: 463–473.

Goldman, Armond S. 2001. Breastfeeding lessons from the past century. *Pediatric Clinics of North America* 48: 23A–25A.

Goldman, Armond S., Chheda, Sadhana, Garofalo, Roberto. 1998. Evolution of immunologic functions of the mammary gland and the postnatal development of immunity. *Pediatric Research* 43: 155–162.

Goldman, S. E., Schneider, H. G. 1987. Menstrual synchrony: Social and personality factors. *Journal of Social Behavior and Personality* 2: 243–250.

Golub, M. S. 2000. Adolescent health and the environment. *Environmental Health Perspectives* 108: 355–362.

Golub, Sharon, ed. 1985. *Lifting the Curse of Menstruation*. New York: Harrington Park Press.

Gomez, Enrique, Ortiz, Victor, Saint-Martin, Blanca, Boeck, Lourdes, Diaz-Sanchez, Vicente, Bourges, Hector. 1993. Hormonal regulation of the secretory IgA (sIgA) system:

Estradiol- and progesterone-induced changes in sIgA in parotid saliva along the menstrual cycle. *American Journal of Reproductive Immunology* 29: 219–223.

Goodman, Catherine Chase. 2003. Intergenerational triads in grandparent-headed families. *Journal of Gerontology: Social Sciences* 58B: S281–S289.

Graafmans, Wilco C., Richardus, Jan Hendrik, Borsboom, Gerard J. J. M., Bakketeig, Leiv, Langhoff-Roos, Jens, Bergsjø, Per, Macfarlane, Alison, Verloove-Vanhorick, S. Pauline, Mackenbach, Johan P., group, EuroNatal Working Group. 2002. Birth weight and perinatal mortality: A comparison of "optimal" birth weight in seven western European countries. *Epidemiology* 13: 569–574.

Graber, Julia A., Lewinsohn, P. M., Seeley, J. R., Brooks Gunn, Jeanne. 1997. Is psychopathology associated with timing of pubertal development? *Journal of American Academy of Child and Adolescent Psychiatry* 36: 1768–1776.

Gracia, C. R., Freeman, E. W. 2004. Acute consequences of the menopausal transition: The rise of common menopausal symptoms. *Endocrinology and Metabolism Clinics of North America* 33: 675–689.

Graham, C. A., McGrew, W. C. 1980. Menstrual synchrony in female undergraduates living on a coeducational campus. *Psychoneuroendocrinology* 5: 245–252.

Grummer-Strawn, Laurence M., Scanlon, Kelley S., Fein, Sara B. 2008. Infant feeding and feeding transitions during the first year of life. *Pediatrics* 122: S36–S42.

Guerneri, Sillvana, Bettio, Daniela, Simoni, G., Brambati, B., Lanzani, A., Fraccaro, M. 1987. Prevalence and distribution of chromosome abnormalities in a sample of first trimester internal abortions. *Human Reproduction* 2: 735–739.

Gurven, Michael, Hill, Kim. 1997. Comment on Hawkes, et al. *Current Anthropology* 38: 566–567.

Gurven, Michael, Kaplan, Hillard, Winking, Jeffrey, Finch, Caleb E., Crimmins, Eileen M. 2008. Aging and inflammation in two epidemiological worlds. *Journal of Gerontology* 3A: 196–199.

Hagen, E. H. 1999. The functions of postpartum depression. *Evolution and Human Behavior* 20: 325–359.

Haig, David. 1993. Genetic conflicts in human pregnancy. *Quarterly Review of Biology* 68: 495–532.

Haig, David. 1999. Genetic conflicts in pregnancy and childhood. In *Evolution in Health and Disease*, ed. S. C. Stearns, pp. 77–90. Oxford, UK: Oxford University Press.

Haig, David. 2008. Intimate relations: Evolutionary conflicts of pregnancy and childhood. In *Evolution in Health and Disease*, ed. Stephen C Stearns, Jacob C. Koella, pp. 65–76. Oxford, UK: Oxford University Press.

Hall, Roberta. 2004. An energetics-based approach to understanding the menstrual cycle and menopause. *Human Nature* 15: 83–89.

Hamosh, Margit. 2001. Bioactive factors in human milk. *Pediatric Clinics of North America* 48: 64–86.

Handa, Victoria L., Pannu, Harpreet K, Siddique, Sohail, Gutman, Robert, VanRooyen, Julia, Cundiff, Geoff. 2003. Architectural differences in the bony pelvis of women with and without pelvic floor disorders. *Obstetrics & Gynecology* 102: 1283–1290.

Hanlon-Lundberg, K. M., Kirby, R. S. 2000. Association of ABO incompatibility with elevation of nucleated red blood cell counts in term neonates. *American Journal of Obstetrics and Gynecology* 183: 1532–1536.

Hannah, Mary E. 2004. Planned elective cesarean section: A reasonable choice for some women? *Canadian Medical Association Journal* 170: 813–814.

Harder, Thomas, Bergmann, Renate, Kallischnigg, Gerd, Plagemann, Andreas. 2005. Duration of breastfeeding and risk of overweight: A meta-analysis. *American Journal of Epidemiology* 162: 397–403.

Harit, D., Faridi, M. M. A., Aggarwal, A., Sharma, S. B. 2008. Lipid profile of term infants on exclusive breastfeeding and mixed feeding: Comparative study. *European Journal of Clinical Nutrition*. 62: 203–209.

Harris, M., Ross, E. B., eds. 1987. *Food and Evolution: Toward a Theory of Human Food Habits*. Philadelphia, PA: Temple University Press.

Harvey, S. M. 1987. Female sexual behavior: Fluctuations during the menstrual cycle. *Journal of Psychosomatic Research* 31: 101–110.

Hasselbalch, H., Engelmann, M. D. M., Ersboll, A. K., Jeppesen, D. L., Fleischer-Michaelsen, K. 1999. Breast-feeding influences thymic size in late infancy. *European Journal of Pediatrics* 158: 964–967.

Hasselbalch, H., Jeppesen, D. L., Engelmann, M. D. M., Michaelsen, K. F., Nielsen, M. B. 1996. Decreased thymus size in formula-fed infants compared with breastfed infants. *Acta Paediatrica* 85: 1029–1032.

Hawkes, Kristen. 1991. Showing off: Tests of an hypothesis about men's foraging goals. *Ethology and Sociobiology* 12: 29–54.

Hawkes, Kristen. 2003. Grandmothers and the evolution of human longevity. *American Journal of Human Biology* 15: 380–400.

Hawkes, K., O'Connell, J. F., Blurton Jones, N. G. 1997. Hadza women's time allocation, offspring provisioning, and the evolution of long postmenopausal life spans. *Current Anthropology* 38: 551–577.

Hays, Warren S. T. 2003. Human pheromones: Have they been demonstrated? *Behavioral Ecology and Sociobiology* 54: 89–97.

Heinig, M. Jane. 2001. Host defense benefits of breastfeeding for the infant. *Pediatric Clinics of North America* 58: 105–123.

Heinig, M. Jane, Dewey, Kathryn G. 1997. Health effects of breast feeding for mothers: A critical review. *Nutrition Research Reviews* 10: 35–56.

Hellin, K., Waller, G. 1992. Mothers' mood and infant feeding: Prediction of problems and practices. *Journal of Reproductive and Infant Psychology* 10: 39–51.

Hill, E. M. 1988. The menstrual cycle and components of human female sexual behaviour. *Journal of Social and Biological Structures* 11: 443–455.

Hill, K., Hurtado, A. M. 1991. The evolution of reproductive senescence and menopause in human females. *Human Nature* 24: 315–350.

Hoath, Steven B., Narendran, Vivek, Visscher, Marty O. 2001. The biology of vernix. *Newborn and Infant Nursing Reviews* 1: 53–58.

Hoath, Steven B., Pickens, W. L., Visscher, Marty O. 2006. The biology of vernix caseosa. *International Journal of Cosmetic Science* 28: 319–333.

Holte, A., Mikkelsen, A. 1991. The menopausal syndrome: A factor analytic replication. *Maturitas* 13: 193–203.

Horne, Rosemary S. C., Franco, Patricia, Adamson, T. Michael, Groswasser, José, Kanh, André. 2002. Effects of body position on sleep and arousal characteristics in infants. *Early Human Development* 69: 25–33.

Hotvedt, Mary. 1991. Sexuality in the later years: The cross-cultural and historical context. In *Human Sexuality: Cross-Cultural Readings*, ed. Brian M. du Toit, pp. 278–297. New York: McGraw-Hill.

Hrdy, S. B. 1999. *Mother Nature: A History of Mothers, Infants, and Natural Selection*. New York: Ballantine

Hrdy, Sarah Blaffer. 2005. Cooperative breeders with an ace in the hole. In *Grandmotherhood*, ed. Eckart Voland, Athanasios Chasiotis, W. Schiefenhövel. New Brunswick, NJ: Rutgers University Press.

Hunter, M. S. 1988. Psychological aspects of the climacteric and postmenopause. In *The Menopause*, ed. J. W. W. Studd, M. I. Whitehead. Oxford, UK: Blackwell Scientific Publications.

Hunter, M. S. 1996. Depression and the menopause. *British Medical Journal* 313: 1217–1218.

Jacobson, Sandra W., Jacobson, Joseph. 2006. Editorial: Breast feeding and intelligence in children. *British Medical Journal* 333: 929–930.

Jarett, L. R. 1984. Psychosocial and biological influences on menstruation: Synchrony, cycle length, and regularity. *Psychoneuroendocrinology* 9: 21–28.

Jasienska Grażnya, Thune, Inger. 2001. Lifestyle, hormones, and risk of breast cancer. *British Medical Journal* 322: 586–587.

Jasienska, G., Ziomkiewicz, A., Ellison, P. T., Lipson, S. F., Thune, I. 2006. Habitual physical activity and estradiol levels in women of reproductive age. *European Journal of Cancer Prevention* 15: 439–445.

Jenni, Oskar G. 2004. Sleep-wake processes play a key role in early infant crying. *Behavioral and Brain Sciences* 27: 464–465.

Johns, T., Duquette, M. 1991. Detoxification and mineral supplementation as functions of geophagy. *American Journal of Clinical Nutrition* 53: 448–456.

Jolly, Alison. 1985. *The Evolution of Primate Behavior*. New York: Macmillan.

Jordan, B. 1997. Authoritative knowledge and its construction. In *Childbirth and Authoritative Knowledge*, ed. Robbie Davis-Floyd, Carol Sargent, pp. 55–79. Berkeley: University of California Press.

Jurmain, Robert, Kilgore, Lynn, Trevathan, Wenda, Ciochon, Russell L. 2008. *Introduction to Physical Anthropology*, 11th ed. Belmont, CA: Thomson Wadsworth.

Kaltiala-Heino, R., Martunnen, M., Rantanen, P., Rimpelä, M. 2003. Early puberty is associated with mental health problems in middle adolescence. *Social Science & Medicine* 57: 1055–1064.

Kaplan H., Hill K., Lancaster J., Hurtado A. M. 2000. A theory of human life history evolution: Diet, intelligence, and longevity. *Evolutionary Anthropology* 9: 156–185.

Kaplan, Hillard S., Robson, Arthur J. 2002. The emergence of humans: The coevolution of intelligence and longevity with intergenerational transfers. *Proceedings of the National Academy of Sciences* 99: 10221–10226.

Karaolis-Danckert, Nadina, Günther, Anke, L. B., Kroke, Anja, Hornberg, Claudia, Buyken, Anette E. 2007. How early dietary factors modify the effect of rapid weight gain in infancy on subsequent body-composition development in term children whose birth weight was appropriate for gestational age 1–3. *American Journal of Clinical Nutrition* 86: 1700–1708.

Kaufert, Patricia A., Lock, Margaret. 1992. "What are women for?": Cultural constructions of menopausal women in Japan and Canada. In *In Her Prime*, ed. Virginia Kerns, Judith K. Brown, pp. 201–219. Urbana: University of Illinois Press.

Kaushic, Charu, Zhou, Fan, Murdin, Andrew D., Wira, Charles R. 2000. Effects of estradiol and progesterone on susceptibility and early immune responses to *Chlamydia trachomatis* infection in the female reproductive tract. *Infection and Immunity* 68: 4207–4216.

Kelleher, Christa M. 2006. The physical challenges of early breastfeeding. *Social Science & Medicine* 63: 2727–2738.

kellymom. 2009. Financial costs of not breastfeeding. http://kellymom.com. Accessed May, 2009.

Kendall-Tackett, Kathleen. 2007. A new paradigm for depression in new mothers: The central role of inflammation and how breastfeeding and anti-inflammatory treatments protect maternal mental health. *International Breastfeeding Journal* 2: 1–14.

Kennedy, G. E. 2005. From the ape's dilemma to the weanling's dilemma: Early weaning and its evolutionary context. *Journal of Human Evolution* 48: 123–145.

Kerns, Virginia, Brown, Judith K., eds. 1992. *In Her Prime*. Urbana: University of Illinois Press.

King, Janet C. 2003. The risk of maternal nutritional depletion and poor outcomes increases in early or closely spaced pregnancies. *Journal of Nutrition* 133: 1732S–1736S.

Kirkwood, T. B. L. 2008. Understanding ageing from an evolutionary perspective. *Journal of Internal Medicine* 263: 117–127.

Kitzinger, Sheila. 2009. Letter from Europe: A quick fix for crying? *Birth* 36: 86–87.

Klaus, M. H., Kennell, J. H. 1976. *Mother-infant Bonding*. St. Louis: Mosby.

Kluger, M. J. 1978. The evolution and adaptive value of fever. *American Scientist* 66: 38–43.

Knipp, Gregory T., Audus, Kenneth L., Soares, Michael J. 1999. Nutrient transport across the placenta. *Advanced Drug Delivery Reviews* 38: 41–58.

Knott, Cheryl. 2001. Female reproductive ecology of the apes. In *Reproductive Ecology and Human Reproduction*, ed. P. T. Ellison, pp. 429–463. New York: Aldine de Gruyter.

Knox, K., Baker, J. C. 2008. Genomic evolution of the placenta using co-option and duplication and divergence. *Genome Research* 18: 695–705.

Kramer, Karen L. 2008. Early sexual maturity among Pumé foragers of Venezuela: Fitness implications of teen motherhood. *American Journal of Physical Anthropology* 136: 338–350.

Kramer K. L., McMillan, G. P. 2006. The effect of labor-saving technology on longitudinal fertility changes. *Current Anthropology* 47: 165–72.

Krogman, W. M. 1951. The scars of human evolution. *Scientific American* 185: 54–57.

Kronborg, Hanne, Væth, Michael. 2009. How are effective breastfeeding technique and pacifier use related to breastfeeding problems and duration? *Birth* 36: 34–42.

Kuh, D., Hardy, R., Rodgers, B., Wadsworth, M. E. J. 2002. Lifetime risk factors for women's psychological distress in midlife. *Social Science and Medicine* 55: 1957–1973.

Kuh, Diana, dos Santos Silva, Isabel, Barrett-Connor, Elizabeth. 2002. Disease trends in women living in established market economies: Evidence of cohort effects during the epidemiological transition. In *A Life-Course Approach to Women's Health*, ed. Diana Kuh, Rebecca Hardy, pp. 347–373. Oxford, UK: Oxford University Press.

Kuzawa, Christopher W. 1998. Adipose tissue in human infancy and childhood: An evolutionary perspective. *American Journal of Physical Anthropology* 107: 177–209.

Kuzawa, Christopher W. 2005. Fetal origins of developmental plasticity: Are fetal cues reliable predictors of future nutritional environments? *American Journal of Human Biology* 17: 5–21.

Kuzawa, Christopher W. 2007. Developmental origins of life history: Growth, productivity, and reproduction. *American Journal of Human Biology* 19: 654–661.

Kuzawa, Christopher W. 2008. The developmental origins of adult health: Intergenerational inertia in adaptation and disease. In *Evolutionary Medicine and Health: New Perspectives*, ed. Wenda R. Trevathan, E. O. Smith, James J. McKenna, pp. 325–349. New York: Oxford University Press.

Labbok, Miriam H. 2001. Effects of breastfeeding on the mother. *Pediatric Clinics of North America* 48: 143–158.

Lack, D., Gibb, J. A., Owen, D. F. 1957. Survival in relation to brood-size in tits. *Proceedings of the Zoological Society of London* 128: 313–326.

Lagercrantz, Hugo, Slotkin, Theodore A. 1986. The "stress" of being born. *Scientific American* 254: 100–107.

Lahdenperä, Mirkka, Lummaa, Virpl, Helle, Samull, Tremblay, Marc, Russell, Andrew F. 2004. Fitness benefits of prolonged post-reproductive lifespan in women. *Nature* 428: 178–181.

Laitman, Jeffrey T. 1984. The anatomy of human speech. *Natural History* 93: 20–27.

Laitman, Jeffrey T. 1986. Comment on McKenna, 1986. *Medical Anthropology* 10: 65–66.

Lancaster, Jane, King, Barbara J. 1992. An evolutionary perspective on menopause. In *In Her Prime*, ed. Virginia Kerns, Judith K. Brown, pp. 7–15. Urbana: University of Illinois Press.

Lancaster, J. B., Lancaster, C. S. 1983. Parental investment: The hominid adaptation. In *How Humans Adapt: A Biocultural Odyssey*, ed. D. J. Ortner, pp. 33–65. Washington, DC: Smithsonian Institution Press.

Lassek, William D., Gaulin, Steven J. C. 2007. Brief communication: Menarche is related to fat distribution. *American Journal of Physical Anthropology* 133: 1147–1151.

Lassek, William D., Gaulin, Steven J. C. 2008. Waist-hip ratio and cognitive ability: Is gluteofemoral fat a privileged store of neurodevelopmental resources? *Evolution and Human Behavior* 29: 26–34.

Law, Malcolm. 2000. Dietary fat and adult diseases and the implications for childhood nutrition: An epidemiologic approach. *American Journal of Clinical Nutrition* 72 (suppl): 1291S–1296S.

Lawrence, Robert M., Lawrence, Ruth A. 2001. Given the benefits of breastfeeding, what contraindications exist? *Pediatric Clinics of North America* 48: 235–251.

Lee, Richard B. 1968. What hunters do for a living, or how to make out on scarce resources. In *Man the Hunter*, ed. Richard B. Lee, Irven deVore, pp. 30–48. Chicago: Aldine.

Lee, Richard B. 1980. Lactation, ovulation, infanticide, and women's work: A study of hunter-gatherer population regulation. In *Biosocial Mechanisms of Population Regulation*, ed. Mark N. Cohen, Roy S. Malpass, Harold G. Klein, pp. 321–348. New Haven: Yale University Press.

Lee, Richard B. 1992. Work, sexuality, and aging among !Kung women. In *In Her Prime*, ed. Virginia Kerns, Judith K. Brown, pp. 36–46. Urbana: University of Illinois Press.

Lee, Ronald D. 2003. Rethinking the evolutionary theory of aging: Transfers, not births, shape senescence in social species. *Proceedings of the National Academy of Sciences* 100: 9637–9642.

Leidy, L. E. 1996. Symptoms of menopause in relation to the timing of reproductive events and past menstrual experience. *American Journal of Human Biology* 8: 761–769.

Leidy, L. 1999. Menopause in evolutionary perspective. In *Evolutionary Medicine*, ed. W. R. Trevathan, E. O. Smith, J. J. McKenna, pp. 407–428. New York: Oxford University Press.

Leonetti, Donna L., Nath, Dilip C., Hemam, Natabar S. 2007. In-law conflict. *Current Anthropology* 48: 861–890.

Lepore, Jill. 2009. Baby food: If breast is best, why are women bottling their milk? *The New Yorker*, January 19, pp. 34–39.

Levine, R. J., Qian, C., Maynard, S. E., Yu, K. F., Epstein, F. H., Karumanchi, S. A., Levine, Richard J., Qian, Cong, Maynard, Sharon E., Yu, Kai F., Epstein, Franklin H., Karumanchi, S. Ananth. 2006. Serum sFlt1 concentration during preeclampsia and mid trimester blood pressure in healthy nulliparous women. *American Journal of Obstetrics and Gynecology* 194: 1034–1041.

Li, M. J., Short, R. 2002. How oestrogen or progesterone might change a woman's susceptibility to HIV-1 infection. *Australian and New Zealand Journal of Obstetrics and Gynaecology* 42: 472–475.

Li, Ruowei, Fein, Sara B., Chen, Jian, Grummer-Strawn, Laurence M. 2008. Why mothers stop breastfeeding: Mothers' self-reported reasons for stopping during the first year. *Pediatrics* 122: S69–S76.

Liestøl, Knut. 1980. Menarcheal age and spontaneous abortion: A causal connection? *American Journal of Epidemiology* 111: 753–758.

Lindsay, Pat. 2001. Cost of NOT breastfeeding. http://patlc.com/cost_not_breastfeeding.htm. Accessed January 5, 2009.

Lipsitt, Lewis, Burns, Barbara. 1986. Comment on McKenna, 1986." vol="10">Lipsitt, Lewis, Burns, Barbara. 1986. Comment on McKenna, 1986. *Medical Anthropology* 10: 66–67.

Lipson, S. F. 2001. Metabolism, maturation and ovarian function. In *Reproductive Ecology and Human Evolution*, ed. P.T. Ellison, pp. 235–244. New York: Aldine de Gruyter.

Lipson, S. F., Ellison, P. T. 1992. Normative study of age variation in salivary progesterone profiles. *Journal of Biosocial Sciences* 24: 233–244.

Liston, W. A. 2003. Rising caesarean section rates: Can evolution and ecology explain some of the difficulties of modern childbirth? *Journal of the Royal Society of Medicine* 96: 559–561.

Livingstone, Frank. 1958. Anthropological implications of sickle cell gene distribution in West Africa. *American Anthropologist* 60: 533–562.

Lock, M. 1993. *Encounters with Aging: Mythologies of Menopause in Japan and North America*. Berkeley: University of California Press.

Lock, Margaret, Kaufert, Patricia A. 2001. Menopause, local biologies, and cultures of aging. *American Journal of Human Biology* 13: 494–504.

Locke, John L., Bogin, B. 2006. Language and life history: A new perspective on the development and evolution of human language. *Behavioral and Brain Sciences* 29: 259–325.

Loisel, Dagan A., Alberts, Susan C., Ober, Carole. 2008. Functional significance of MHC variation in mate choice, reproductive outcome, and disease risk. In *Evolution in Health and Disease*, ed. Stephen C. Stearns, Jacob C. Koella, pp. 95–108. Oxford, UK: Oxford University Press.

Lovejoy, C. Owen. 2005. The natural history of human gait and posture: Part I, spine and pelvis. *Gait and Posture* 21: 95–112.

Luckett, P. W. 1974. Reproductive development and evolution of the placenta in primates. *Contributions to Primatology* 3: 142–234.

Mace, Ruth. 1998. The coevolution of human fertility and wealth inheritance strategies. *Philosophical Transactions of the Royal Society of London. Series B, Biological Sciences* 353: 389–397.

Mace, Ruth. 2000. Evolutionary ecology of human life history. *Animal Behaviour* 59: 1–10.

Mace, Ruth, Sear, Rebecca. 2005. Are humans cooperative breeders? In *Grandmotherhood*, ed. Eckart Voland, Athanasios Chasiotis, W. Schiefenhövel, pp. 143–159. New Brunswick, NJ: Rutgers University Press.

Madrigal, Lorena, Meléndez-Obando, Mauricio. 2008. Grandmothers' longevity negatively affects daughters' fertility. *American Journal of Physical Anthropology* 136: 223–229.

Maes, Michael, Linb, Ai-hua, Ombelete, Willem, Stevense, Karolien, Kenisf, Gunter, De Jonghg, Raf, Coxh, John, Bosmansf, Eugène. 2000. Immune activation in the early puerperium is related to postpartum anxiety and depressive symptoms. *Psychoneuroendocrinology* 25: 121–137.

Marchini, G., Berggren, V., Djilali-Merzoug, R., Hansson, L-O. 2000. The birth process initiates an acute phase reaction in the fetus-newborn infant. *Acta Paediatrica* 89: 1082–1086.

Marchini, G., Lindow, S., Brismar, H., Stabi, B., Berggren, V., Ulfgren, A-K., Lonne-Rahm, S., Agerberth, B., Gudmundsson, G. H. 2002. The newborn infant is protected by an innate antimicrobial barrier: Peptide antibiotics are present in the skin and vernix caseosa. *British Journal of Dermatology* 147: 1127–1134.

Martin, Emily 1987. *The Woman in the Body: A Cultural Analysis of Reproduction*. Boston: Beacon Press.

Martin, Emily. 1988. Premenstrual syndrome: Discipline, work and anger in late industrial societies. In *Blood Magic: The Anthropology of Menstruation*, ed. Thomas Buckley, Alma Gottlieb, pp. 161–181. Berkeley: University of California Press.

Martin, R. D. 2007. Evolution of human reproduction. *Yearbook of Physical Anthropology* 50: 59–84.

Martin, Richard M., Ebrahim, Shah, Griffin, Maura, Smith, George Davey, Nicolaides, Andrew N., Georgiou, Niki, Watson, Simone, Frankel, Stephen, Holly, Jeff M. P., Gunnell, David. 2005. Breastfeeding and atherosclerosis. *Arteriosclerosis, Thrombosis, and Vascular Biology* 25: 1482–1488.

Martin, Richard M., Goodall, Sarah H., Gunnell, David, Smith, George Davey. 2007. Breast feeding in infancy and social mobility: 60-year follow-up of the Boyd Orr cohort. *Archives of Disease in Childhood* 92: 317–321.

Martin, Richard M., Gunnell, David, Smith, George Davey. 2005. Breastfeeding in infancy and blood pressure in later life: Systematic review and meta-analysis. *American Journal of Epidemiology* 161: 15–26.

Martin, Richard M., Smith, G. Davey, Mangtani, P., Frankel, S., Gunnell, D. 2002. Association between breast feeding and growth: The Boyd-Orr cohort study. *Archives of Disease in Childhood Fetal and Neonatal Edition* 87: F193.

Martorell, R. 1989. Body size, adaptation, and function. *Human Organization* 48: 15–20.

Matsunaga, E. 1959. Selection in ABO polymorphism in Japanese populations. *American Journal of Human Genetics* 11: 405–413.

Matteo, S., Rissman, E. F. 1984. Increased sexual activity during the midcycle portion of the human menstrual cycle. *Hormones and Behavior* 18: 249–255.

Matthews, Karen A., Wing, Rena R., Kuller, Lewis H., Meilahn, Elaine N., et al. 1990. Influences of natural menopause on psychological characteristics and symptoms of middle-aged healthy women. *Journal of Consulting and Clinical Psychology* 58: 345–351.

Mayer, Peter. 1982. Evolutionary advantages of menopause. *Human Ecology* 10: 477–494.

Maynard Smith, J., Barker, D. J. P., Finch, C. E., Kardia, S. L. R., Eaton, S. B., Kirkwood, T. B. L., LeGrand, E. K., Nesse, R. M., Williams, G. C., Partridge, L. 1999. The evolution of non-infectious and degenerative disease. In *Evolution in Health and Disease*, ed. S. C. Stearns, pp. 267–272. Oxford, UK: Oxford University Press.

McClintock, M. K. 1971. Menstrual synchrony and suppression. *Nature* 229: 244–245.

McCoy, Norma, Cutler, Winnifred, Davidson, Julian M. 1985. Relationships among sexual behavior, hot flashes, and hormone levels in perimenopausal women. *Archives of Sexual Behavior* 14: 385–394.

McDade, Thomas W. 2005. The ecologies of human immune function. *Annual Review of Anthropology* 34: 495–521.

McDade, Thomas W., Beck, M. A., Kuzawa, Christopher W., Adair, Linda S. 2001. Prenatal undernutrition and postnatal growth are associated with adolescent thymic function. *Journal of Nutrition* 131: 1225–1231.

McDade, Thomas W., Worthman, Carol M. 1998. The weanling's dilemma reconsidered: A biocultural analysis of breastfeeding ecology. *Developmental and Behavioral Pediatrics* 19: 286–299.

McDade, Thomas W., Worthman, Carol M. 1999. Evolutionary process and ecology of human immune function. *American Journal of Human Biology* 11: 705–717.

McEwan, B. S. 2003. Mood disorders and allostatic load. *Biological Psychiatry* 54: 200–207.

McKenna, James J. 1986. An anthropological perspective on the sudden infant death syndrome (SIDS): The role of parental breathing cues and speech breathing adaptations. *Medical Anthropology* 10: 9–53.

McKenna, James. J. 1996. Sudden infant death syndrome in cross-cultural perspective: Is infant-parent cosleeping protective? *Annual Review of Anthropology* 25: 201–216.

McKenna, James J., Ball, Helen L., Gettler, Lee T. 2007. Mother-infant cosleeping, breastfeeding, and sudden infant death syndrome: What biological anthropology has discovered about normal infant sleep and pediatric sleep medicine. *Yearbook of Physical Anthropology* 50: 133–161.

McKenna, James J., McDade, Thomas W. 2005. Why babies should never sleep alone: A review of the co-sleeping controversy in relation to SIDS, bedsharing and breast feeding. *Paediatric Respiratory Reviews* 6: 134–152.

McKenna, James J., Mosko, S., Richard, C. 1999. Breastfeeding and mother-infant co-sleeping in relation to SIDS prevention. In *Evolutionary Medicine*, ed. W. R. Trevathan, E. O. Smith, J. J. McKenna, pp. 53–74. New York: Oxford University Press.

Michel, Sven C. A., Rake, Annett, Treiber, Karl, Seifert, Burkhardt, Chaoui, Rabih, Huch, Renate, Marincek, Borut, Kubic-Huch, Rahel A. 2002. MR Obstetric pelvimetry: Effect of birthing position on pelvic bony dimensions. *American Journal of Roentgenology* 179: 1063–1067.

Mikiel-Kostyra, K., Mazur, J., Bołtruszko, I. 2002. Effect of early skin-to-skin contact after delivery on duration of breastfeeding: A prospective study. *Acta Pædiatrica* 91: 1301–1306.

Mills, Margaret E. 2007. Craving more than food: The implications of pica in pregnancy. *Nursing for Women's Health* 11: 266–273.

Milne, Eugene M. G. 2006. When does human ageing begin? *Mechanisms of Ageing and Development* 127: 290–297.

Mintz, Jessica. 2009. Facebook draws fire from nursing mothers. In *The New Mexican*, October 6, pp. A1, A5.

Miranda, S. B. 1970. Visual abilities and pattern preference of premature and full-term neonates. *Journal of Experimental Child Psychology* 10: 189–205.

Mitchell, E. S., Woods, N. F. 1996. Symptom experiences of midlife women: Observations from the Seattle midlife women's health study. *Maturitas* 25: 1–10.

Moalem, Sharon. 2007. *Survival of the Sickest*. New York: William Morrow.

Moerman, Marquisa LaVelle. 1982. Growth of the birth canal in adolescent girls. *American Journal of Obstetrics and Gynecology* 143: 528–532.

Moffitt, Terrie E., Caspi, Avshalom, Belsky, Jay, Silva, Phil A. 1992. Childhood experience and the onset of menarche: A test of a sociobiological model. *Child Development* 63: 47–58.

Montagu, Ashley. 1961. Neonatal and infant immaturity in man. *Journal of the American Medical Association* 178: 56–57.

Montagu, Ashley. 1964. *Life before Birth*. New York: New American Library.

Montagu, Ashley. 1978. *Touching: The Human Significance of the Skin*, 2nd ed. New York: Harper and Row.

Moore, B., Kombe, H. 1991. Climacteric symptoms in a Tanzanian community. *Maturitas* 13: 229–234.

Morgan, W. P. 1982. Psychological effects of exercise. *Behavioral Medicine Update* 4: 25–30.

Muehlenbein, M. P., Bribiescas, Richard. G. 2005. Testosterone-mediated immune functions and male life histories. *American Journal of Human Biology* 17: 227–258.

Murray, L., Fiori-Cowley, A., Hooper, P. 1996. The impact of postnatal depression and associated adversity on early mother-infant interactions and later infant outcomes. *Child Development* 67: 2512–2526.

Musil, Carol M. 2000. Health of grandmothers as caregivers: A ten month follow-up. *Journal of Women & Aging* 12: 129–145.

Nagata, C., Shimizu, S., Takami, R., Hayashi, M., Takeda, N., Yasuda, K. 1999. Hot flushes and other menopausal symptoms in relation to soy product intake in Japanese women. *Journal of the International Menopause Society* 2: 6–12.

Nagata, C., Takatsuka, N., Inaba, S., Kawakami, N., Shimizu, H. 1998. Association of diet and other lifestyle with onset of menopause in Japanese women. *Maturitas* 29: 105–113.

Nathanielsz, P. W. 2001. *The Prenatal Prescription*. New York: HarperCollins.

Nepomnaschy, P. A., Welch, K. M. A., McConnell, D., Low, B. S., Strassmann, Beverly I., England, B. G. 2006. Cortisol levels and very early pregnancy loss in humans. *Proceedings of the National Academy of Sciences* 103: 3938–3942.

Nesse, R. M. 1991. What good is feeling bad? *The Sciences*, November/December: 30–37.

Nesse, R. M. 2001. Medicine's missing basic science. *New Physician* 50: 8–10.

Nesse, Randolph M., Schiffman, Joshua D. 2003. Evolutionary biology in the medical school curriculum. *Bioscience* 53: 585–587.

Nesse, Randolph M., Stearns, Stephen C. 2008. The great opportunity: Evolutionary applications to medicine and public health. *Evolutionary Applications* 1: 28–48.

Nesse, Randolph M., Stearns, S. C., Omenn, G. S. 2006. Medicine needs evolution. *Science* 311: 1071.

Nesse, R. M., Williams, G. C. 1994. *Why We Get Sick—The New Science of Darwinian Medicine.* New York: Times Books.

Nesse, R. M., Williams, G. C. 1997. Evolutionary biology in the medical curriculum: What every physician should know. *Bioscience* 47: 664–666.

Neville, Margaret C. 2001. Anatomy and physiology of lactation. *Pediatric Clinics of North America* 48: 13–34.

Neville, Margaret C., Morton, Jane, Umemura, Shinobu. 2001. Lactogenesis: The transition from pregnancy to lactation. *Pediatric Clinics of North America* 48: 35–52.

Nicol-Smith, L. 1996. Causality, menopause, and depression: A critical review of the literature. *BMJ* 313: 1229–1232.

Nikolaou, D., Gilling-Smith, C. 2004. Early ovarian ageing: Are women with polycystic ovaries protected? *Human Reproduction* 19: 2175–2179.

Nishida, Toshisada. 1997. Comment on Hawkes et al. *Current Anthropology* 38: 568–569.

Núñez-de la Mora, Alejandra, Bentley, Gillian R. 2008. Early life effects on reproductive function. In *Evolutionary Medicine and Health: New Perspectives*, ed. W. R. Trevathan, E. O. Smith, J. J. McKenna, pp. 149–168. New York: Oxford University Press.

Nygaard, Ingrid, Barber, Matthew D., Burgio, Kathryn L., Kenton, Kimberly, Meikle, Susan, Schaffer, Joseph, Spino, Cathie, Whitehead, William W., Wu, Jennifer, Brody, Debra J., Network, Pelvic Floor Disorders. 2008. Prevalence of symptomatic pelvic floor disorders in US women. *Journal of the American Medical Association* 300: 1311–1316.

Oates, M. R., Cox, J. L., Neema, S., Asten, P., Glangeaud-Freudenthal, N., Figuieredo, B., Gorman, L. L., Hacking, S., Hirst, E., Kammerer, M. H., Klier, C. M., Seneviratne, G., Smith, M. A., Suntter-Dallay, A. -L., Valoriani, V., Wickberg, B., Yoshida, K., Group, TCS-PND. 2004. Postnatal depression across countries and cultures: A qualitative study. *British Journal of Psychiatry* 184: s10–s16.

Ober, C., Elias, S. G., Kostyu, D. D., Hauck, W. W. 1992. Decreased fecundability in Hutterite couples sharing HLA-DR. *American Journal of Human Genetics* 50: 6–14.

Obermeyer, Carla Makhlouf, Castle, Sarah. 1997. Back to nature? Historical and cross-cultural perspectives on barriers to optimal breastfeeding. *Medical Anthropology* 17: 39–63.

Oddy, Wendy H. 2004. A review of the effects of breastfeeding on respiratory infections, atopy, and childhood asthma. *Journal of Asthma* 41: 605–621.

Oddy, Wendy H., Sheriff, Jill L., de Klerk, Nicholas H., Kendall, Garth E., Sly, Peter D., Beilin, Lawrence J., Blake, Kevin B., Landau, Louis I., Stanley, Fiona J. 2004. The relation of breastfeeding and body mass index to asthma and atopy in children: A prospective cohort study to age 6 years. *American Journal of Public Health* 94: 1531–1537.

Odent, Michel. 2003. *Birth and Breastfeeding.* East Sussex, UK: Clairview.

Okonofua, F. E., Lawal, A., Bamgbose, J. K. 1990. Features of menopause and menopausal age in Nigerian women. *International Journal of Gynecology and Obstetrics* 31: 341–345.

Olshansky, S. Jay, Carnes, Bruce A., Butler, Robert N. 2001. If humans were built to last. *Scientific American* 284: 50–55.

O'Rahilly, R., Muller, F. 1998. Developmental morphology of the embryo and fetus. In *Cambridge Encyclopedia of Human Growth and Development*, ed. S. J. Ulijaszek, F. E. Johnston, M. A. Preece, pp. 161–162. Cambridge, UK: Cambridge University Press.

Panter-Brick, Catherine, Pollard, Tessa M. 1999. Work and hormonal variation in subsistence and industrial contexts. In *Hormones, Health, and Behavior*, ed. C. Panter-Brick, C. M. Worthman, pp. 139–183. Cambridge, UK: Cambridge University Press.

Papousek, H., Papousek, M. 1990. Excessive infant crying and intuitive parental care: Buffering support and its failures in parent-infant interaction. *Early Child Development and Care* 65: 117–125.

Parkin, D. Maxwell, Bray, Freddie, Ferlay, J., Pisani, Paola. 2005. Global cancer statistics, 2002. *CA: A Cancer Journal for Clinicians* 55: 74–108.

Parlee, M. B. 1990. Integrating biological and social scientific research in menopause. *Annals of the New York Academy of Sciences* 592: 379–389.

Paul, Andreas. 2005. Primate predispositions for human grandmaternal behavior. In *Grandmotherhood*, ed. Eckart Voland, Athanasios Chasiotis, W. Schiefenhövel, pp. 21–58. New Brunswick, NJ: Rutgers University Press.

Pavelka, Mary S. M., Fedigan, Linda M. 1991. Menopause: A comparative life history perspective. *Yearbook of Physical Anthropology* 34: 13–38.

Pavelka, Mary M., Fedigan, Linda M., Zohar, Sandra. 2002. Availability and adaptive value of reproductive and postreproductive Japanese macaque mothers and grandmothers. *Animal Behaviour* 64: 407–414.

Pawlowski, B. 1999. Loss of estrus and concealed ovulation in human evolution. The case against the sexual selection hypothesis. *Current Anthropology* 40: 257–275.

Pawlowski, Broguslaw, Grabarczyk, Marzena. 2003. Center of body mass and the evolution of female body shape. *American Journal of Human Biology* 15: 144–150.

Peacock, N. 1990. Comparative and cross-cultural approaches to the study of human female reproductive failure. In *Primate Life History and Evolution*, ed. C. J. DeRousseau, pp. 195–220. New York: Wiley-Liss.

Peccei, Jocelyn Scott. 1995. A hypothesis for the origin and evolution of menopause. *Maturitas* 21: 83–89.

Peccei, Jocelyn Scott. 2001. A critique of the grandmother hypotheses: Old and new. *American Journal of Human Biology* 13: 434–452.

Perls, Thomas T., Fretts, Ruth C. 2001. The evolution of menopause and the human life span. *Annals of Human Biology* 28: 237–245.

Pettitt, David J., Forman, Michele R., Hanson, Robert L., Knowler, William C., Bennett, Peter H. 1997. Breastfeeding and incidence of non-insulin-dependent diabetes mellitus in Pima Indians. *Lancet* 350: 166–168.

Phipps, Maureen G., Sowers, MaryFran. 2002. Defining early adolescent childbearing. *American Journal of Public Health* 92: 125–128.

Picciano, Mary Frances. 2001. Nutrient composition of human milk. *Pediatric Clinics of North America* 48: 53–67.

Pike, IL. 2000. The nutritional consequences of pregnancy sickness: A critique of a hypothesis. *Human Nature* 11: 207–232.

Piperata, Barbara Anne. 2008. Forty days and forty nights: A biocultural perspective on postpartum practices in the Amazon. *Social Science & Medicine* 67: 1094–1103.

Pollard, Tessa M. 2008. *Western Diseases: An Evolutionary Perspective*. Cambridge, UK: Cambridge University Press.

Pollard, T., Unwin, N. 2008. Impaired reproductive function in women in Western and "westernizing" populations: An evolutionary approach. In *Evolutionary Medicine and*

Health: New Perspectives, ed. W. R. Trevathan, E. O. Smith, J. J. McKenna, pp. 169–180. New York: Oxford University Press.

Portmann, Adolf. 1990. *A Zoologist Looks at Humankind.* New York: Columbia University Press.

Power, M. L., Tardif, S. D. 2005. Maternal nutrition and metabolic control of pregnancy. In *Birth, Distress and Disease: Placental-Brain Interactions*, ed. M. L. Power, J. Schulkin, pp. 88–112. Cambridge, UK: Cambridge University Press.

Prakash, I. J., Murthy, Vinoda N. 1982. Menopausal symptoms in Indian women. *Personality Study and Group Behavior* 2: 54–58.

Preti, G., Cutler, W. B., Garcia, C. T. 1986. Human axillary secretions influence women's menstrual cycles: The role of donor extract of females. *Hormones and Behavior* 20: 474–482.

Profet, M. 1992. Pregnancy sickness as adaptation: A deterrent to maternal ingestion of teratogens. In *The Adapted Mind: Evolutionary Psychology and the Generation of Culture*, ed. J. H. Barkow, L Cosmides, J Tooby, pp. 327–365. New York: Oxford University Press.

Profet, M. 1993. Menstruation as a defense against pathogens transported by sperm. *Quarterly Review of Biology* 68: 335–386.

Quadagno, D. M., Shubeita, H. E., Deck, J., Francoeur, D. 1981. Influence of male social contacts, exercise and all-female living conditions on the menstrual cycle. *Psychoneuroendocrinology* 6: 239–244.

Rafael, Dana. 1973. *The Tender Gift: Breastfeeding.* New York: Schocken Books.

Read, A. W., Prendiville, W. J., Dawes, V. P., Stanley, F. J. 1994. Cesarean section and operative vaginal delivery in low risk primiparous women, Western Australia. *American Journal of Public Health* 84: 37–42.

Redman, C. W., Sargent, I. L. 2005. Latest advances in understanding preeclampsia. *Science* 308: 1592–1594.

Reiber, Chris. 2008. An evolutionary model of premenstrual syndrome. *Medical Hypotheses* 70: 1058–1065.

Reynolds, Ann. 2001. Breastfeeding and brain development. *Pediatric Clinics of North America* 48: 159–171.

Riordan, Jan. 1997. The cost of not breastfeeding: A commentary. *Journal of Human Lactation* 13: 93–97.

Robillard, P.-Y., Dekker, G. A., Hulsey, T. C. 2002. Evolutionary adaptations to pre-eclampsia/eclampsia in humans: Low fecundability rate, loss of oestrus, prohibitions of incest and systematic polyandry. *American Journal of Reproductive Immunology* 47: 104–111.

Robillard, Pierre-Yves, Dekker, Gustaaf, Chaouat, Gérard, Chaline, Jean, Hulsey, Thomas C. 2008. Possible role of eclampsia/preeclampsia in evolution of human reproduction. In *Evolutionary Medicine and Health: New Perspectives*, ed. W. R. Trevathan, E. O. Smith, J. J. McKenna, pp. 216–225. New York: Oxford University Press.

Robinson, G. 1996. Cross cultural perspectives on menopause. *Journal of Nervous and Mental Disease* 184: 453–458.

Robson, S. L., van Schaik, C. P., Hawkes, K. 2006. The derived features of human life history. In *The Evolution of Human Life History*, ed. K. Hawkes, R. R. Paine. Santa Fe, NM: School of American Research Press.

Rosenberg, Karen R. 1992. The evolution of modern human childbirth. *Yearbook of Physical Anthropology* 35: 89–124.

Rosenberg, Karen R. 2004. Living longer: Information revolution, population expansion, and modern human origins. *Proceedings of the National Academy of Sciences* 101: 10847–10848.

Rosenberg, Karen R., Trevathan, Wenda R. 1996. Bipedalism and human birth: The obstetrical dilemma revisited. *Evolutionary Anthropology* 4: 161–168.

Rosenberg, Karen R., Trevathan, Wenda R. 2001. The evolution of human birth. *Scientific American* 285: 72–77.

Rosenberg, Karen R., Trevathan, Wenda R. 2002. Birth, obstetrics and human evolution. *BJOG: An International Journal of Obstetrics and Gynaecology* 109: 1199–1206.

Rosenblatt, Karin A., Thomas, David B., The WHO Collaborative Study of Neoplasia and Steroid Contraceptives. 1995. Prolonged lactation and endometrial cancer. *International Journal of Epidemiology* 24: 499–503.

Ross, R. K., Coetzee, G. A., Richardt, J., Skinner, E., Henderson, B. E. 1995. Does the racial-ethnic variation in prostate cancer risk have a hormonal basis? *Cancer Causes and Control* 75: 1778–1782.

Roy, Robert P. 2003. A Darwinian view of obstructed labor. *Obstetrics & Gynecology* 101: 397–401.

Russell, M. J., Switz, G. M., Thompson, K. 1980. Olfactory influences on the human menstrual cycle. *Pharmacology, Biochemistry and Behavior* 13: 737–738.

Ryan, Alan Se., Zhou, Wenjun. 2006. Lower breastfeeding rates persist among the special supplemental nutrition program for Women, Infants, and Children participants, 1978–2003. *Pediatrics* 117: 1136–1146.

Salk, Lee. 1960. The effects of the normal heartbeat sound on the behavior of the newborn infant: Implications for mental health. *World Mental Health* 12: 168–175.

Sanabria, Emilia. 2009. The politics of menstrual suppression in Brazil. *Anthropology News* 50: 6–7.

Sands, Roberta G., Goldberg-Glen, Robin, Thornton, Pamela L. 2005. Factors associated with the positive well-being of grandparents caring for their grandchildren. *Journal of Gerontological Social Work* 45: 65–82.

Schank, Jeffrey C. 2000. Menstrual-cycle variability and measurement: Further cause for doubt. *Psychoneuroendocrinology* 25: 837–847.

Schank, Jeffrey C. 2006. Do human menstrual-cycle pheromones exist? *Human Nature* 17: 433–447.

Schimpf, Megan, Tulikangas, Paul. 2005. Evolution of the female pelvis and relationships to pelvic organ prolapse. *International Urogynecology Journal* 16: 315–320.

Schindler, A. E. 2002. Mood disorders and other menopause related problems. *Maturitas* 41: S1.

Schmidt, P. J., Haq, N., Rubinow, D. R. 2004. A longitudinal evaluation of the relationship between reproductive status and mood in perimenopausal women. *American Journal of Psychiatry* 161: 2238–2244.

Scholl, T. O., Hediger, M. L., Schall, J. I., Khoo, C., Fischer, R. L. 1994. Maternal growth during pregnancy and the competition for nutrients. *American Journal of Clinical Nutrition* 60: 183–188.

Schön, Regine A. 2007. Natural parenting: Back to basics in infant care. *Evolutionary Psychology* 5: 102–183.

Schooling, Mary, Kuh, Diana. 2002. A life course perspective on women's health behaviours. In *A Life Course Approach to Women's Health*, ed. Diana Kuh, Rebecca Hardy, pp. 279–303. Oxford, UK: Oxford University Press.

Schuitemaker, N., Van Roosmalen, J., Dekker, G., Van Dongen, P., Van Geijn, H., Gravenhorst, J. B. 1997. Maternal mortality after cesarean section in The Netherlands. *Acta Obstetricia et Gynecologica Scandinavica* 76: 332–334.

Schultz, A. H. 1949. Sex differences in the pelves of primates. *American Journal of Physical Anthropology* 7: 401–424.

Scott, K. D., Klaus, P. H., Klaus, M. H. 1999. The obstetrical and postpartum benefits of continuous support during childbirth. *Journal of Women's Health and Gender-Based Medicine* 8: 1257–1264.

Sear, Rebecca, Mace, Ruth. 2008. Who keeps children alive? A review of the effects of kin on child survival. *Evolution and Human Behavior* 29: 1–18.

Seckl, J. R., Drake, A. J., Holmes, M. C. 2005. Prenatal glucocorticoids and the programming of adult diseases. In *Birth, Distress, and Disease*, ed. M. L. Power, J. Schulkin. Cambridge, UK: Cambridge University Press.

Seckler, D. 1982. Small but healthy? A basic hypothesis in the theory, measurement and policy of malnutrition. In *Newer Concepts in Nutrition and Their Implications for Policy*, ed. P. V. Sukhatme, pp. 127–137. Pune, India: Maharashtra Association for the Cultivation of Science Research Institute.

Sellen, Daniel W. 2006. Lactation, complementary feeding, and human life history. In *The Evolution of Human Life History*, ed. Kristen Hawkes, Richard R. Paine, pp. 155–196. Santa Fe, NM: SAR Press.

Sellen, Daniel W. 2007. Evolution of infant and young child feeding: Implications for contemporary public health. *Annual Review of Nutrition* 27: 123–148.

Shanley, Daryl P., Kirkwood, Thomas B. L. 2001. Evolution of the human menopause. *BioEssays* 23: 282–287.

Shanley, Daryl P., Sear, Rebecca, Mace, Ruth, Kirkwood, Thomas B. L. 2007. Testing evolutionary theories of menopause. *Proceedings of the Royal Society B, Biological Sciences* 274: 2943–2949.

Sheiner, Eyal, Levy, Amalia, Katz, Miriam, Mazor, Moshe. 2004. Short stature—an independent risk factor for Cesarean delivery. *European Journal of Obstetrics & Gynecology and Reproductive Biology* 120: 175–178.

Short, Roger V. 1976. The evolution of human reproduction. *Proceedings of the Royal Society of London Series B - Biological Sciences* 195: 3–24.

Shostak, Marjorie. 1981. *Nisa*. New York: Vintage Books.

Shubin, N. 2008. *Your Inner Fish*. New York: Random House.

Sievert, Lynette Leidy. 2006. *Menopause: A Biocultural Perspective*. New Brunswick, NJ: Rutgers University Press.

Sievert, Lynette Leidy. 2008. Should women menstruate? An evolutionary perspective on menstrual-suppressing oral contraceptives. In *Evolutionary Medicine and Health: New Perspectives*, ed. Wenda R. Trevathan, E. O. Smith, James J. McKenna, pp. 181–195. New York: Oxford University Press.

Sievert, Lynette Leidy, Waddle, Diane, Canali, Kristophor. 2001. Marital status and age at natural menopause: Considering pheromonal influence. *American Journal of Human Biology* 13: 479–485.

Signore, C., Mills, J. L., Qian, C., Yu, K., Lam, C., Epstein, F. H., Karumanchi, S. A., Levine, R. J., Signore, Caroline, Mills, James L., Qian, Cong, Yu, Kai, Lam, Chun, Epstein, Franklin H., Karumanchi, S. Ananth, Levine, Richard J. 2006. Circulating angiogenic factors and placental abruption. *Obstetrics and Gynecology* 108: 338–344.

Simkin, Penny. 2003. Maternal positions and pelves revisited. *Birth* 30: 130–132.

Simmons, David. 1997. NIDDM and breastfeeding. *The Lancet* 350: 157–158.

Singh, Gurcharan, Archana, G. 2008. Unraveling the mystery of vernix caseosa. *Indian Journal of Dermatology* 53: 54–60.

Small, M. F. 1998. *Our Babies Ourselves—How Biology and Culture Shape the Way We Parent.* New York: Doubleday Dell.

Smith, B. H. 1991. Dental development and the evolution of life history in Hominidae. *American Journal of Physical Anthropology* 86: 157–174.

Smith, Stephen M., Baskin, Gary B., Marx, Preston A. 2000. Estrogen protects against vaginal transmission of simian immunodeficiency virus. *Journal of Infectious Diseases* 182: 708–715.

Soltis, Joseph. 2004. The signal functions of early infant crying. *Behavioral and Brain Sciences* 27: 443–409.

Sorensen, H. J., Mortensen, E. L., Reinisch, J. M., Mednick, S. A. 2005. Breastfeeding as risk of schizophrenia in the Copenhagen Perinatal Cohort. *Acta Psychiartica Scandinavica* 112: 26–29.

Sorenson Jamison, Cheryl, Jamison, Paul, Cornell, Laurel L. 2005. Human female longevity. In *Grandmotherhood*, ed. Eckart Voland, Athanasios Chasiotis, W. Schiefenhövel, pp. 99–117. New Brunswick, NJ: Rutgers University Press.

Stanford, J. L., Hartge, P., Brinton, L. A., Hoove, R. N., Brookmeyer, R. 1987. Factors influencing the age at natural menopause. *Journal of Chronic Disease* 40: 995–1002.

Stearns, S. C. 1992. *The Evolution of Life Histories.* Oxford, UK: Oxford University Press.

Stearns, S. C., Ebert, D. 2001. Evolution in health and disease: Work in progress. *Quarterly Review of Biology* 76: 417–432.

Stearns, S. C., Koella, J. 1986. The evolution of phenotypic plasticity in life-history traits: predictions for reaction norms for age and size at maturity. *Evolution* 40: 893-913

Steer, P. J. 2006. Prematurity or immaturity? *BJOG: An International Journal of Obstetrics and Gynaecology* 113: 136–138.

Stewart, D. B. 1984. The pelvis as a passageway. I. Evolution and adaptations. *British Journal of Obstetrics & Gynaecology* 91: 611–617.

Stoller, M. 1995. *The Obstetric Pelvis and Mechanism of Labor in Nonhuman Primates.* Chicago: University of Chicago.

Strassman, B. I. 1996. The evolution of endometrial cycles and menstruation. *Quarterly Review of Biology* 71: 181–220.

Strassmann, Beverly I. 1997. The biology of menstruation in *Homo sapiens*: Total lifetime menses, fecundity, and nonsynchrony in a natural-fertility population. *Current Anthropology* 38: 123–129.

Strassmann, Beverly I. 1999a. Menstrual cycling and breast cancer: An evolutionary perspective. *Journal of Women's Health* 8: 193–202.

Strassmann, Beverly I. 1999b. Menstrual synchrony pheromones: Cause for doubt. *Human Reproduction* 14: 579–580.

Stuebe, Alison. 2007. Duration of lactation and maternal metabolism at 3 years postpartum. *American Journal of Obstetrics and Gynecology* 197: S128 – S128.

Stuebe, Alison M., Rich-Edwards, Janet W., Willett, Walter C., Manson, JoAnn E., Michels, Karin B. 2006. Duration of lactation and incidence of type 2 diabetes. *Obstetrical & Gynecological Survey* 61: 232–233.

Sukwatana, P., Meekhangvan, J., Tamrongterakul, T., Tanapat, Y., Asavarait, S., Boonjitrpimon, P. 1991. Menopausal symptoms among Thai women in Bangkok. *Maturitas* 13: 217–228.

Surbey, Michele K. 1990. Family composition, stress and the timing of human menarche. In *Socioendocrinology of Primate Reproduction*, ed. Toni E. Ziegler, Fred B. Bercovitch. New York: Wiley-Liss.

Swain, James E., Tasgin, Esra, Mayes, Linda C., Feldman, Ruth, Constable, R. Todd, Leckman, James F. 2008. Maternal brain response to own baby-cry is affected by cesarean section delivery. *Child Psychology and Psychiatry* 49: 1042–1052.

Tansirikongkol, Anyarporn, Wickett, R. Randall, Visscher, M. O., Hoath, Steven B. 2007. Effect of vernix caseosa on the penetration of chymotryptic enzyme: Potential role in epidermal barrier development. *Pediatric Research* 62: 49–53.

Tardif, S. D., Power, M. L., Layne, D., Smucny, D., Ziegler, T. 2004. Energy restriction initiated at different gestational age has varying effects on maternal weight gain and pregnancy outcome in common marmoset monkeys (Callithrix jacchus). *British Journal of Nutrition* 92: 841–849.

Taylor, Diana. 2006. From "It's all in your head " to "Taking back the month": Premenstrual syndrome (PMS) research and the contributions of the Society for Menstrual Cycle Research. *Sex Roles* 54: 377–391.

Temmermana, M., Verstraelena, H., Martensa, G., Bekaert, A. 2004. Delayed childbearing and maternal mortality. *European Journal of Obstetrics & Gynecology and Reproductive Biology* 114: 19–22.

Thomas, Frédéric, Renaud, François, Benefice, Eric, de Meeüs, Thierry, Guegan, Jean-François. 2001. International variability of ages at menarche and menopause: Patterns and main determinants. *Human Biology* 73: 271–290.

Thompson, Melissa Emery. 2005. Reproductive endocrinology of wild female chimpanzees (*Pan troglodytes schweinfurthii*): Methodological considerations and the role of hormones in sex and conception. *American Journal of Primatology* 67: 137–158.

Thompson, Melissa Emery, Jones, James H., Pusey, Anne E., Brewer-Marsden, Stella, Goodall, Jane, Marsden, David, Matsuzawa, Tetsuro, Nishida, Toshisada, Reynolds, Vernon, Sugiyama, Yukimaru, Wrangham, Richard W. 2007. Aging and fertility patterns in wild chimpanzees provide insights into the evolution of menopause. *Current Biology* 47: 2150–2156.

Thompson, Melissa Emery, Wilson, M. L., Gobbo, G., Muller, M. N., Pusey, A. E. 2008. Hyperprogesteronemia in response to Vitex fischeri consumption in wild chimpanzees (*Pan troglodytes schweinfurthii*). *American Journal of Primatology* 70: 1064–1071.

Thompson, Nicholas S., Olson, Carolyn, Dessureau, Brian. 1996. Babies' cries: Who's listening? Who's being fooled? *Social Research* 63: 763–784.

Tollin, M., Bergsson, G., Kai-Larsen, Y., Lengqvist, J., Sjovall, J., Griffiths, W., Skulandottir, G. V., Haraldsson, A., Jornvall, H., Gudmundsson, G. H., Agerberth, B. 2005. Vernix caseosa as a multi-component defence system based on polypeptides, lipids, and their interactions. *Cellular and Molecular Life Sciences* 62: 2390–2399.

Tracer, David P. 2002. Somatic versus reproductive energy allocation in Papua New Guinea: Life history theory and public health policy. *American Journal of Human Biology* 14: 621–626.

Tracer, David P. 2009. Infant carrying and prewalking locomotor development: Proximate and evolutionary perspectives. Paper presented at the American Association of Physical Anthropologists, Chicago.

Trevathan, Wenda R. 1981. Maternal touch at first contact with the newborn infant. *Developmental Psychobiology* 14: 549–558.

Trevathan, Wenda R. 1982. Maternal lateral preference at first contact with her newborn infant. *Birth* 9: 85–90.

Trevathan, Wenda R. 1987. *Human Birth: An Evolutionary Perspective*. New York: Aldine de Gruyter.

Trevathan, Wenda R. 1988. Fetal emergence patterns in evolutionary perspective. *American Anthropologist* 90: 19–26.

Trevathan, Wenda R. 1999. Evolutionary obstetrics. In *Evolutionary Medicine*, ed. W. R. Trevathan, E. O. Smith, J. J. McKenna, pp. 407–427. New York: Oxford University Press.

Trevathan, Wenda R. 2007. Evolutionary medicine. *Annual Review of Anthropology* 36: 139–154.

Trevathan, Wenda R., Burleson, Mary H., Gregory, W. Larry. 1993. No evidence for menstrual synchrony in lesbian couples. *Psychoneuroendocrinology* 18: 425–435.

Trevathan, Wenda R., McKenna, James J. 1994. Evolutionary environments of human birth and infancy: Insights to apply to contemporary life. *Children's Environments* 11: 88–104.

Trevathan, Wenda R., Rosenberg, Karen R. 2000. The shoulders follow the head: Postcranial constraints on human childbirth. *Journal of Human Evolution* 39: 583–586.

Trevathan, Wenda R., Smith, E. O., McKenna, James J. 2008a. Introduction and overview of evolutionary medicine. In *Evolutionary Medicine and Health: New Perspectives*, ed. Wenda R. Trevathan, E. O. Smith, James J. McKenna, pp. 1–54. New York: Oxford University Press.

Trevathan, Wenda R., Smith, E. O., McKenna, James J., eds. 2008b. *Evolutionary Medicine and Health: New Perspectives*. New York: Oxford University Press.

Trzonkowski, P., Mysliwska, J., Lucaszuk, K, Szmit, E., Bryl, E., Mysliwski, A. 2001. Luteal phase of the menstrual cycle in young healthy women is associated with decline in interleukin 2 levels. *Hormone and Metabolic Research* 33: 348–353.

Uvnas-Moberg, Kerstin. 1989. The gastrointestinal tract in growth and reproduction. *Scientific American* 261: 78–83.

Uvnas-Moberg, Kerstin. 1996. Neuroendocrinology of the mother-child interaction. *Trends in Endocrinology & Metabolism* 7: 126–131.

Uvnas-Moberg, Kerstin. 1998. Oxytocin may mediate the benefits of positive social interaction and emotions. *Psychoneuroendocrinology* 23: 819–835.

Valeggia, Claudia, Ellison, Peter T. 2003. Lactational amenorrhoea in well-nourished Toba women of Formosa, Argentina. *Journal of Biosocial Sciences* 36: 573–595.

van Schaik, Carel P., Barrickman, Nancy, Bastian, Meredith L., Krakauer, Elissa B., van Noordwijk, Maria A. 2006. Primate life histories and the role of brains. In *The Evolution of Human Life History*, ed. Kristen Hawkes, Richard R. Paine. Santa Fe, NM: SAR Press.

Varendi, Heili, Porter, Richard H., Winberg, Jan. 2002. The effect of labor on olfactory exposure learning within the first postnatal hour. *Behavioral Neuroscience* 116: 206–211.

Varner, Michael W., Fraser, Alison M., Hunter, Cheri Y., Corneli, Patrice S., Ward, Ryk H. 1996. The intergenerational predisposition to operative delivery. *Obstetrics and Gynecology* 87: 905–911.

Vatuk, Sylvia. 1992. Sexuality and the middle-aged woman in South Asia. In *In Her Prime*, ed. Virginia Kerns, Judith K. Brown, pp. 155–170. Urbana: University of Illinois Press.

Venkatesha, S., Toporsian, M., Lam, C., Hanai, J., Mammoto, T., Kim, Y. M., Bdolah, Y., Lim, K. H., Yuan, H. T., Libermann, T. A., Stillman, I. E., Roberts, D., D'Amore, P. A., Epstein, F. H., Sellke, F. W., Romero, R., Sukhatme, V. P., Letarte, M., Karumanchi, S. A. 2006. Soluble endoglin contributes to the pathogenesis of preeclampsia. *Nature Medicine* 12: 642–649.

Vihko, R., Apter, D. 1984. Endocrine characteristics of adolescent menstrual cycles: Impact of early menarche. *Journal of Steroid Biochemistry* 20: 231–236.

Visscher, Marty O., Narendran, Vivek, Pickens, William L., LaRuffa, Angela A., Meinzen-Derr, Jareen, Allen, Kathleen, Hoath, Steven B. 2005. Vernix caseosa in neonatal adaptation. *Journal of Perinatology* 25: 440–446.

Vitzthum, Virginia J. 1997. Flexibility and paradox: The nature of adaptation in human reproduction. In *The Evolving Female*, ed. Mary Ellen Morbeck, Alison Galloway, Adrienne L. Zihlman, pp. 242–258. Princeton: Princeton University Press.

Vitzthum, Virginia. 2001. Why not so great is still good enough. In *Reproductive Ecology and Human Evolution*, ed. P. T. Ellison, pp. 179–202. New York: Aldine de Gruyter.

Vitzthum, Virginia. 2008. Evolutionary models of women's reproductive functioning. *Annual Review of Anthropology* 37: 53–73.

Vitzhum, V. 2009. The ecology and evolutionary endocrinology of reproduction in the human female. *Yearbook of Physical Anthropology* 52: 95–136.

Vitzthum, Virginia, Bentley, G. R., Spielvogel, H., Caceres, E., Thornburg, J., Jones, L., Shore, S., Hodges, K. R., Chatterton, R. J. 2002. Salivary progesterone levels and rate of ovulation are significantly lower in poorer than in better-off urban-dwelling Bolivian women. *Reproduction* 17: 1906–1913.

Vitzthum, Virginia, Spielvogel, H. 2003. Epidemiological transitions, reproductive health, and the Flexible Response Model. *Economics and Human Biology* 2003: 223–242.

Vitzthum, Virginia, Spielvogel, Hilde, Caceres, Esperanza, Gaines, Julia. 2000. Menstrual patterns and fecundity among non-lactating and lactating cycling women in rural highland Bolivia: Implications for contraceptive choice. *Contraception* 62: 181–187.

Vitzthum, Virginia, Spielvogel, Hilde, Caceres, Esperanza, Miller, Aaron. 2001. Vaginal bleeding patterns among rural highland Bolivian women: Relationship to fecundity and fetal loss. *Contraception* 64: 319–325.

Vitzthum, Virginia, Spielvogel, H., Thornburg, J. 2004. Interpopulational differences in progesterone levels during conception and implantation in humans. *Proceedings of the National Academy of Sciences of the United States of America* 101: 1443–1448.

Vitzthum, Virginia, Spielvogel, H., Thornburg, J. 2006. A prospective study of early pregnancy loss in humans. *Fertility & Sterility* 86: 373–379.

Voland, Eckart, Beise, Jan. 2002. Opposite effects of maternal and paternal grandmothers on infant survival in historical Krummhörn. *Behavioral Ecology and Sociobiology* 52: 435–443.

Voland, Eckart, Chasiotis, Athanasios, Schiefenhövel, Wulf, eds. 2005. *Grandmotherhood: The Evolutionary Significance of the Second Half of Female Life*. New Brunswick, NJ: Rutgers University Press.

Walker, Margaret L., Herndon, James G. 2008. Menopause in nonhuman primates? *Biology of Reproduction* 79: 398–406.

Walker, Robert S., Gurven, Michael, Burger, Oskar, Hamilton, Marcus J. 2008. The trade-off between number and size of offspring in humans and other primates. *Proceedings of the Royal Society B, Biological Sciences* 275: 827–833.

Wallace, Jacqueline M., Aitken, Raymond P., Milne, John S., Hay, William W. 2004. Nutritionally mediated placental growth restriction in the growing adolescent: Consequences for the fetus. *Biology of Reproduction* 71: 1055–1062.

Wall-Scheffler, Cara M., Geiger, K., Steudel-Numbers, Karen L. 2007. Infant carrying: The role of increased locomotory costs in early tool development. *American Journal of Physical Anthropology* 133: 841–846.

Walrath, Dana. 2003. Rethinking pelvic typologies and the human birth mechanism. *Current Anthropology* 44: 5–31.

Walrath, Dana. 2006. Gender, genes, and the evolution of human birth. In *Feminist Anthropology: Past, Present, and Future*, ed. P. L. Geller, M. K. Stockett, pp. 55–69. Philadelphia: University of Pennsylvania Press.

Walsh, Joseph A. 2008. Evolution and the cesarean section rate. *American Biology Teacher* 70: 401–404.

Wang, Thomas W., Apgar, Barbara S. 1998. Exercise during pregnancy. *American Family Physician* 57: 1846–1852.

Wasser, Samuel K., Isenberg, David Y. 1986. Reproductive failure among women: Pathology or adaptation? *Journal of Psychosomatic Obstetrics and Gynaecology* 5: 153–175.

Waterhouse, J. A. H., Hogben, L. 1947. Incompatibility of mother and foetus with respect to the isoagglutinogen A and its antibody. *British Journal of Social Medicine* 1: 1–17.

Waterlow, J. C., Thomson, A. M. 1979. Observations on the adequacy of breast-feeding. *The Lancet* 314: 238–242.

Wei, G., Greaver, L. B., Marson, S. M., Herndon, C. H., Rogers, J. 2008. Postpartum depression: Racial differences and ethnic disparities in a tri-racial and bi-ethnic population. *Maternal & Child Health Journal* 12: 699–707.

Weinstock, M. 2001. Alterations induced by gestational stress in brain morphology and behaviour of the offspring. *Progress in Neurobiology* 65: 427–451.

Weiss, Robin Elise. 2008. Why labor is good for babies. In *About.com: Pregnancy and childbirth*. Accessed November 25, 2008.

Weller, Leonard, Weller, Aron, Koresh-Kamin, Hagit, Ben-Shoshan, Rivi. 1999. Menstrual synchrony in a sample of working women. *Psychoneuroendocrinology* 24: 449–459.

Westphal, Sylvia P. 2004. The best skin cream you ever wore. *New Scientist* 181: 40.

Whitcome, Katherine K., Shapiro, Liza J., Leiberman, Daniel E. 2007. Fetal load and the evolution of lumbar lordosis in bipedal hominins. *Nature Genetics* 450: 1075–1078.

White, Emily, Velentgas, P., Mandelson, M. T., Lehman, C. D., Elmore, J. G., Porter, P., Yasui, Y., Taplin. S. H.1998. Variation in mammographic breast density by time in menstrual cycle among women aged 40–49 years. *Journal of the National Cancer Institute* 90: 906–910.

Whiteman, M. K., Staopoli, C. A., Langenberg, P. W., McCarter, R. J., Kjerulff, K. H., Flaws, J. A. 2003. Smoking, body mass, and hot flashes in midlife women. *Obstetrics and Gynecology* 101: 264–272.

Whitten P., Naftolin F. 1998. Reproductive actions of phytoestrogens. *Baillieres Clinical Endocrinology and Metabolism* 12: 667–90.

Whitten, P. L. 1999. Diet, hormones and health: An evolutionary-ecological perspective. In *Hormones, Health, and Behavior: A Socio-ecological and Lifespan Perspective*, ed. Catherine Panter-Brick, C. M. Worthman, pp. 210–242. Cambridge, UK: Cambridge University Press.

WHO Multicentre Growth Reference Study Group. 2006. WHO Child Growth Standards based on length/height, weight and age. *Acta Pædiatrica* Suppl 450: 76–85.

Wichstrom, Lars. 2000. Predictors of adolescent suicide attempts: A nationally representative longitudinal study of Norwegian adolescents. *Journal of the American Academy of Child and Adolescent Psychiatry* 39: 603–610.

Wilbur, JoEllen, Dan, Alice, Hedricks, Cynthia, Holm, Karyn. 1990. The relationship among menopausal status, menopausal symptoms, and physical activity in midlife women. *Family and Community Health* 13: 67–78.

Wiley, Andrea S. 2008. Cow's milk consumption and health: An evolutionary perspective. In *Evolutionary Medicine and Health: New Perspectives*, ed. Wenda Trevathan, E. O. Smith, James J. McKenna, pp. 116–133. New York: Oxford University Press.

Wiley, Andrea S., Allen, John S. 2009. *Medical Anthropology: A Biocultural Approach*. New York: Oxford University Press.

Wiley, Andrea S., Katz, Solomon H. 1998. Geophagy in pregnancy: A test of a hypothesis. *Current Anthropology* 39: 532–545.

Williams, G. C. 1957. Pleiotropy, natural selection, and the evolution of senescence. *Evolution* 11: 398–411.

Williams, G. C., Nesse, R. M. 1991. The dawn of Darwinian medicine. *Quarterly Review of Biology* 66: 1–22.

Williams, Lawrence. 2008. Aging cebidae. *Interdisciplinary Topics in Gerontology* 36: 49–61.

Williams, Leanne M., Brown, Kerri J., Palmer, Donna, Liddell, Belinda J., Kemp, Andrew H., Olivieri, Gloria, Peduto, Anthony, Gordon, Evian. 2006. The mellow years? Neural basis of improving emotional stability over age. *Journal of Neuroscience* 26: 6422–6430.

Wilson, H. C. 1992. A critical review of menstrual synchrony research. *Psychoneuroendocrinology* 17: 565–591.

Wilson, H. C., Hildebrandt Kiefhaber, S., Gravel, V. 1991. Two studies of menstrual synchrony: Negative results. *Psychoneuroendocrinology* 16: 353–359.

Winberg, Jan. 2005. Mother and newborn baby: Mutual regulation of physiology and behavior—a selective review. *Developmental Psychobiology* 47: 217–229.

Windsor, Timothy D., Anstey, Kaarin J., Rodgers, Bryan. 2008. Volunteering and psychological well-being among young-old adults: How much is too much? *Gerontologist* 48: 59–70.

Winnicott, D. W. 1964. *The Child, the Family, and the Outside World*. Reading, MA: Addison-Wesley.

Wolf, Jacqueline H. 2006. What feminists can do for breastfeeding and breastfeeding can do for feminists. *Signs: Journal of Women in Culture and Society* 31: 397–424.

Wolf, Joan B. 2007. Is breast really best? Risk and total motherhood in the National Breastfeeding Awareness campaign. *Journal of Health Politics, Policy and Law* 32: 595–636.

Wood, J. W. 1994. *Dynamics of Human Reproduction: Biology, Biometry, Demography*. New York: Aldine.

World Health Organization. 2006. *Pregnancy, Childbirth, Postpartum and Newborn Care: A Guide for Essential Practice.* Geneva: Department of Reproductive Health and Research (RHR), World Health Organization.

World Health Organization. 2008. Data and Statistics. http://www.who.int/research/en/. Accessed November 10, 2008.

Worthman, Carol M. 1995. Hormones, sex, and gender. *Annual Review of Anthropology* 24: 593–616.

Worthman, Carol M. 1998. Adolescence in the Pacific: A biosocial view. In *Adolescence in Pacific Island Societies,* ed. G. Herdt, S. C. Leavitt, pp. 27–52. Pittsburgh, PA: University of Pittsburgh Press.

Worthman, Carol M. 1999. Evolutionary perspectives on the onset of puberty. In *Evolutionary Medicine,* ed. W. R. Trevathan, E. O. Smith, J. J. McKenna, pp. 135–164. New York: Oxford University Press.

Worthman, Carol M. 2008. After dark: The evolutionary ecology of human sleep. In *Evolutionary Medicine and Health: New Perspectives,* ed. Wenda R. Trevathan, E. O. Smith, James J. McKenna, pp. 291–313. New York: Oxford University Press.

Worthman, Carol M., Melby, M. 2002. Toward a comparative developmental ecology of human sleep. In *Adolescent Sleep Patterns: Biological, Social, and Psychological Influences,* ed. M. A. Carskadon, pp. 69–117. New York: Cambridge University Press.

Worthman, Carol M., Stallings, J. F. 1994. Measurement of gonadotropins in dried blood spots. *Clinical Chemistry* 40: 448–453.

Worthman, Carol M., Stallings, J. F. 1997. Hormone measures in finger-prick blood spot samples: New field methods for reproductive endocrinology. *American Journal of Physical Anthropology* 104: 1–21.

Wright, A. L. 1983. A cross cultural comparison of menopausal symptoms. *Medical Anthropology* 7: 20–35.

Wright, Anne L. 2001. The rise of breastfeeding in the United States. *Pediatric Clinics of North America* 48: 1–12.

Wyshak, Grace. 1983. Age at menarche and unsuccessful pregnancy outcome. *Annals of Human Biology* 10: 69–73.

Wyshak, G., Frisch, R. E. 2000. Breast cancer among former college athletes compared to non-athletes: A 15-year follow-up. *British Journal of Cancer* 82: 726–30.

Yang, Zhengwei, Schank, Jeffrey C. 2006. Women do not synchronize their menstrual cycles. *Human Nature* 17: 433–477.

Young, T. K., Martens, P. J., Taback, S.P., Sellers, E. A. C., Dean, H. J., Cheang, M., Flett, B. 2002. Type 2 diabetes mellitus in children: Prenatal and early infancy risk factors among native Canadians. *Archives of Pediatric & Adolescent Medicine* 156: 651–655.

Zasloff, Michael. 2003. Vernix, the newborn, and innate defense. *Pediatric Research* 53: 203–204.

Zheng T., Duan L., Liu Y., Zhang B., Wang Y., Chen Y., Zhang Y., Owens P. H. 2000. Lactation reduces breast cancer risk in Shandong Province, China. *American Journal of Epidemiology.* 152(12): 1129–1135.

Ziomkiewicz, Anna. 2006. Menstrual synchrony: Fact or artifact? *Human Nature* 17: 419–432.

Zuk, Marlene. 1997. Darwinian medicine dawning in a feminist light. In *Feminism and Evolutionary Biology,* ed. Patricia Adair Gowaty, pp. 417–430. New York: Chapman and Hall.

Index

Note: Page numbers followed by "*f*" and "*t*" denote figures and tables, respectively.